Notes on Contributors

Gianluca Biggi, Responsible Management Research Center, Department of Economics & Management, University of Pisa, Italy.

John R. Bryson, Department of Strategy and International Business, Birmingham Business School, University of Birmingham, UK.

Alex Coad, Waseda Business School, Waseda University, Tokyo, Japan.

Mirko Dal Maso, UNEP DTU Partnership; Department of Technology, Management and Economics; Technical University of Denmark; Copenhagen; Denmark.

Martin Fougère, Department of Management and Organisation, Hanken School of Economics, Helsinki, Finland.

Emma C. Gardner, Department of Strategy and International Business, Birmingham Business School, University of Birmingham, UK.

Elisa Giuliani, Responsible Management Research Center, Department of Economics & Management, University of Pisa, Italy.

Ulrich Elmer Hansen, UNEP DTU Partnership; Department of Technology, Management and Economics; Technical University of Denmark; Copenhagen; Denmark.

Maureen McKelvey, Institute of Innovation and Entrepreneurship; Department of Economy and Society; School of Business, Economics and Law; University of Gothenburg; Sweden.

Eija Meriläinen, Department of Marketing, Hanken School of Economics, Helsinki, Finland.

Paul Nightingale, SPRU, University of Sussex, Falmer, UK.

Ivan Nygaard, UNEP DTU Partnership; Department of Technology, Management and Economics; Technical University of Denmark; Copenhagen; Denmark.

Rögnvaldur J. Saemundsson, Institute of Innovation and Entrepreneurship; Department of Economy and Society; School of Business, Economics and Law; University of Gothenburg; Sweden. Innovation Management Department, Halmstad University, Sweden. Department of Industrial Engineering, University of Iceland, Reykjavik, Iceland.

Jack Stilgoe, Department of Science & Technology Studies, University College London, UK.

Antonio Vezzani, Department of Economics, Roma Tre University, Italy.

Notes on Contributors

The Dark Side of Innovation

Edited by
Alex Coad, Paul Nightingale, Jack Stilgoe and Antonio Vezzani

Routledge
Taylor & Francis Group
LONDON AND NEW YORK

First published 2022
by Routledge
2 Park Square, Milton Park, Abingdon, Oxon, OX14 4RN

and by Routledge
605 Third Avenue, New York, NY 10158

Routledge is an imprint of the Taylor & Francis Group, an informa business

British Library Cataloguing-in-Publication Data
A catalogue record for this book is available from the British Library

ISBN13: 978-1-032-19418-9 (hbk)
ISBN13: 978-1-032-19422-6 (pbk)
ISBN13: 978-1-003-25908-4 (ebk)

DOI: 10.4324/9781003259084

Typeset in Minion Pro
by codeMantra

Publisher's Note
The publisher accepts responsibility for any inconsistencies that may have arisen during the conversion of this book from journal articles to book chapters, namely the inclusion of journal terminology.

Contents

Citation Information

The chapters in this book were originally published in the journal *Industry and Innovation*, volume 28, issue 1 (2021). When citing this material, please use the original page numbering for each article, as follows:

Chapter 1
Editorial: the dark side of innovation
Alex Coad, Paul Nightingale, Jack Stilgoe and Antonio Vezzani
Industry and Innovation, volume 28, issue 1 (2021) pp. 102–112

Chapter 2
The noxious consequences of innovation: what do we know?
Gianluca Biggi and Elisa Giuliani
Industry and Innovation, volume 28, issue 1 (2021) pp. 19–41

Chapter 3
The dark side of the sun: solar e-waste and environmental upgrading in the off-grid solar PV value chain
Ulrich Elmer Hansen, Ivan Nygaard and Mirko Dal Maso
Industry and Innovation, volume 28, issue 1 (2021) pp. 58–78

Chapter 4
The dark side of the industrialisation of accountancy: innovation, commoditization, colonization and competitiveness
Emma C. Gardner and John R. Bryson
Industry and Innovation, volume 28, issue 1 (2021) pp. 42–57

Chapter 5
The grey zones of technological innovation: negative unintended consequences as a counterbalance to novelty
Maureen McKelvey and Rögnvaldur J. Saemundsson
Industry and Innovation, volume 28, issue 1 (2021) pp. 79–101

Chapter 6

Exposing three dark sides of social innovation through critical perspectives on resilience
Martin Fougère and Eija Meriläinen
Industry and Innovation, volume 28, issue 1 (2021) pp. 1–18

Notes on Contributors

Gianluca Biggi, Responsible Management Research Center, Department of Economics & Management, University of Pisa, Italy.

John R. Bryson, Department of Strategy and International Business, Birmingham Business School, University of Birmingham, UK.

Alex Coad, Waseda Business School, Waseda University, Tokyo, Japan.

Mirko Dal Maso, UNEP DTU Partnership; Department of Technology, Management and Economics; Technical University of Denmark; Copenhagen; Denmark.

Martin Fougère, Department of Management and Organisation, Hanken School of Economics, Helsinki, Finland.

Emma C. Gardner, Department of Strategy and International Business, Birmingham Business School, University of Birmingham, UK.

Elisa Giuliani, Responsible Management Research Center, Department of Economics & Management, University of Pisa, Italy.

Ulrich Elmer Hansen, UNEP DTU Partnership; Department of Technology, Management and Economics; Technical University of Denmark; Copenhagen; Denmark.

Maureen McKelvey, Institute of Innovation and Entrepreneurship; Department of Economy and Society; School of Business, Economics and Law; University of Gothenburg; Sweden.

Eija Meriläinen, Department of Marketing, Hanken School of Economics, Helsinki, Finland.

Paul Nightingale, SPRU, University of Sussex, Falmer, UK.

Ivan Nygaard, UNEP DTU Partnership; Department of Technology, Management and Economics; Technical University of Denmark; Copenhagen; Denmark.

Rögnvaldur J. Saemundsson, Institute of Innovation and Entrepreneurship; Department of Economy and Society; School of Business, Economics and Law; University of Gothenburg; Sweden. Innovation Management Department, Halmstad University, Sweden. Department of Industrial Engineering, University of Iceland, Reykjavik, Iceland.

Jack Stilgoe, Department of Science & Technology Studies, University College London, UK.

Antonio Vezzani, Department of Economics, Roma Tre University, Italy.

The dark side of innovation

Alex Coad, Paul Nightingale, Jack Stilgoe and Antonio Vezzani

ABSTRACT
We provide a broad discussion of the dark side of innovation, before introducing the papers of the special issue. We start with a critical reply to optimists, complementing the list of indicators showing steady human progress with a list of indicators that show sustained deterioration (largely due to innovation). We then outline some relevant dimensions of harmful innovation, before distinguishing between the types of harm brought on by innovation. We conclude with an overview of the SI papers.

1. Introduction

1.1. From optimism to pessimism

For writers such as Steven Pinker (for example, in his book 'Enlightenment Now'), the human condition is steadily improving, and this improvement can be picked up across a range of indicators: fewer homicides, fewer infant deaths, longer life expectancies, improving human rights, increasing literacy, increasing average education, increases in mean (and also median) wealth, fewer people in extreme poverty (less than 1 USD/ day), etc.

However, while not denying significant progress in some areas, particularly the reduction in violence, cherry-picking data to support a comforting Whiggish conclusion is unlikely to generate robust evidence.

To show that the picture is more nuanced, it is worth compiling an alternative, complementary list of dimensions where public health and environmental wellbeing are deteriorating, and highlighting that many of these are probably due to innovation: for example, the appearance and rise of lung cancer (it was relatively unheard of, even among doctors, by 1900: Proctor 2012); the problematic increase of allergies such as asthma (which are somehow conspicuously absent in Amish communities, Holbreich et al. 2012); the rapid increase of anaphylactic shock (Lee et al. 2017); the prolonged decrease in human sperm counts (Levine et al. 2017); the positive relation between new cancer cases and per capita income (which cannot be explained by improved detection potential or longer lives: Luzzati, Parenti, and Rughi 2018); the rise of childhood obesity (Ebbeling, Pawlak, and Ludwig 2002); the global burdens posed by air

pollution (Akimoto 2003), the collapse of insect populations (Kunin 2019); pervasiveness of fluorinated chemicals in the environment (Lim 2019), plastic waste accumulating in the oceans (Cressey 2016); the 'annihilation' of vertebrate populations in the context of the current human-induced mass extinction event (Ceballos, Ehrlich, and Dirzo 2017); the increasing share of Earth's surface made desolate and uninhabitable by human activity (e.g. Chernobyl, Fukushima, polluted battlefields such as Verdun in France (Bausinger, Bonnaire, and Preuß 2007; Cooper 2018), the Union Carbide disaster area in Bhopal with continuing groundwater contamination, perhaps also including landfill sites, nuclear waste disposal areas, former mining sites, landmine fields); and so on.

Rather than cherry picking data to support a preconceived view that innovation is either always good or always bad, we should seek to better understand the nature of innovation (in order to better design innovation policy), and hence there is a need for a more nuanced discussion that rearticulates an old point: innovation has a direction as well as a rate of change (Nelson 1962). Innovation can have good and bad effects, and those positive and negative outcomes are typically unevenly distributed. Choices about innovation are therefore complex and often contested, and the selection environment that weeds out the 'bad' innovations is not something that can be taken for granted. Nor can it be assumed in a pluralistic society, let alone world, that definitions of what counts as good or bad are universally or even widely shared. It would be concerning if people who occupy important positions in society overlook the fact that the definitions of good and bad are reflected in the selection environment for innovation, and think that innovation produces good outcomes naturally.

In recent years, policymakers have seemingly neglected that innovation has a direction, and usually take it for granted that 'innovation is good' and hence, 'more innovation is better'. For example, when they directly compare the performance of countries using indicators such as R&D spending or patenting, more is assumed to be better. Understanding the downsides of innovation, and how those downsides are distributed, might therefore help inform better policy. This editorial, and the special issue, seeks to provide a bit more balance by giving space to suggestions that innovation is not always a force for good.

To do so, we discuss some dimensions we deem useful to conceptualise the link between innovation and its possible harmful effects (section 2); we present a (non-exhaustive) list of conceptually distinct types of harms (section 3); we then discuss three broad unanswered questions from a more practical to a more theoretical perspective (section 4) and, finally; we conclude presenting the papers of this special issue.

2. Dimensions of harmful innovation

This section sets out a (non-exhaustive) number of dimensions according to which innovation may be harmful. These dimensions can be seen as continuums rather than binary distinctions. Future research could add further dimensions to the ones mentioned here or could find an alternative arrangement of dimensions for sorting different types of harmful innovations.

2.1. Issues of scale: explorations by lead users vs the dependence of mass consumers

Small problems due to one-off explorations are generally accepted to be benign and acceptable. For example, small scale pollution from a pioneering scientific experiment is probably acceptable to most people. But if this scales up through the everyday habits of millions of consumers worldwide, harmful effects can accumulate and interact, and new problems can appear. The example of congestion and pollution is a classical case used by economists to illustrate that (public) goods have a capacity beyond which interference (or cumulative) effects become noticeable to then increase disproportionally (Rothenberg 1970). Similar effects apply in many other circumstances and political decisions about the impact of a technology may mistakenly be made by focusing on the initial circumstances, rather than the problems that emerge when the technology diffuses and is used at scale. In this vein, the adoption of new products and technologies can cause problems due to rising demand for key inputs, like the rapid growth in the demand for cobalt (a key input for lithium-ion batteries) causing social problems such as corruption, environmental pollution, extreme poverty and child labour (Conca 2020).

2.2. End-of-product-life considerations

Another dark side of innovation relates to end-of-product-life considerations, that are predictable from the start, but often ignored until the end. Examples include the disposal of nuclear waste, mining clean-up operations, and heavy metal leakage from the recycling of electronic waste. With new technologies, and especially technologies with long working lives, these end of life costs may be uncertain. In some instances, a new technology can reduce these costs, but in many others rising social expectations and more stringent environmental and health regulations can substantially increase costs over time.

Recycling can reduce the environmental impact of waste. However, if new materials such as plastics are relatively cheap, recycling can be prohibitively expensive compared to landfill or burning. In one example, from the UK, Westminster council sent 82% of all household waste – including that put in recycling bins – for incineration in 2017/18 (Franklin-Wallis 2019). Disposal problems are exacerbated in the case of planned obsolescence, where products have deliberately shortened lifetimes, or are designed such that they are impossible or difficult to repair.[1] For example, Apple has been repeatedly accused of deliberately slowing its iPhones in order to increase the sales of new products, and restricts the repair of Apple products even in the case of fairly simple repairs (Svensson et al. 2018).

Despite the usual focus on the benefits of renewable energy, Hansen and Nygaard (this issue) provide evidence on the dangers. New technologies for solar panels cause problems of heavy metal disposal, in particular for lead, which is a potent neurotoxin. Many Sub-Saharan African nations lack capabilities to dispose of waste, and there are uncertainties about the longer-term environmental performance of solar panels. The recycling of hazardous materials is often done by the poor who lack equipment, protection, and political representation. Innovations that address one kind of pollution can create new problems, that fall on different people, and these costs should not be ignored, even if the overall improvement is substantial.

The dark side of innovation

Alex Coad, Paul Nightingale, Jack Stilgoe and Antonio Vezzani

ABSTRACT
We provide a broad discussion of the dark side of innovation, before introducing the papers of the special issue. We start with a critical reply to optimists, complementing the list of indicators showing steady human progress with a list of indicators that show sustained deterioration (largely due to innovation). We then outline some relevant dimensions of harmful innovation, before distinguishing between the types of harm brought on by innovation. We conclude with an overview of the SI papers.

1. Introduction

1.1. From optimism to pessimism

For writers such as Steven Pinker (for example, in his book 'Enlightenment Now'), the human condition is steadily improving, and this improvement can be picked up across a range of indicators: fewer homicides, fewer infant deaths, longer life expectancies, improving human rights, increasing literacy, increasing average education, increases in mean (and also median) wealth, fewer people in extreme poverty (less than 1 USD/day), etc.

However, while not denying significant progress in some areas, particularly the reduction in violence, cherry-picking data to support a comforting Whiggish conclusion is unlikely to generate robust evidence.

To show that the picture is more nuanced, it is worth compiling an alternative, complementary list of dimensions where public health and environmental wellbeing are deteriorating, and highlighting that many of these are probably due to innovation: for example, the appearance and rise of lung cancer (it was relatively unheard of, even among doctors, by 1900: Proctor 2012); the problematic increase of allergies such as asthma (which are somehow conspicuously absent in Amish communities, Holbreich et al. 2012); the rapid increase of anaphylactic shock (Lee et al. 2017); the prolonged decrease in human sperm counts (Levine et al. 2017); the positive relation between new cancer cases and per capita income (which cannot be explained by improved detection potential or longer lives: Luzzati, Parenti, and Rughi 2018); the rise of childhood obesity (Ebbeling, Pawlak, and Ludwig 2002); the global burdens posed by air

2015). Many innovations in the food industry may be harmful. For example, artificial sweeteners are often introduced to reduce the amount of sugar in food. However, they can bring their own problems. Aspartame, for example, is an artificial sweetener whose long-term use has been linked to cancer in mice (Soffritti et al. 2007). 'Processed food' is a synonym for unhealthy food in many food guides.

Even more serious than innovation to escape regulation is innovation to deceive it. This is the case of Volkswagen, who intentionally programmed its diesel engines to give misleading measurements of NOx emissions during regulatory testing (see Jasanoff 2016).

2.5. Sharing the upsides and downsides of innovation

Another issue is related to the 'North-South' perspective. Rich countries develop new innovations that are applied across the world, then the rich countries notice that these innovations are toxic or harmful, and either move on to better alternatives or develop infrastructures to deal with the waste or simply export their waste (e.g. electronic waste). Poor countries, on the other hand, may not have the institutional structures in place to enable them to contain the problems.

Most academic research on innovation focuses on its production in R&D intensive sectors in rich nations. But this is a very misleading view of innovation. For most people, in most countries, for most of history, innovation is not something that was developed where they live to address their problems. Innovation is, instead, something that happens in other places, is adapted to the institutions and culture of those places and comes from those places. Its arrival can be very disruptive because it may not necessarily fit with the local environment, and the changes that are required can be experienced as an imposition.

In this spirit, Hansen and Nygaard (this issue) discuss the problem of solar waste in Sub-Saharan Africa. There is an interesting issue for public policy and normative social science about how the downsides of innovation are socially distributed both internationally and within nations.

3. Types of harm

This section discusses various types of harm from innovation focusing on the aggregated level of harm to society (rather than harm to single individuals) because this is the level where policy decisions are taken.

3.1. Public health risks

While large part of the innovation literature is focused on the private returns from innovation, its impact often goes beyond the producer-consumer relationship, affecting the public sphere. The dichotomy between private and social benefits (costs) may prove useful not only to understand the underinvestment in R&D, but also to understand the overinvestment in specific technologies.

Some public health dangers are not known until long after the innovation's introduction; some were known by firms, some of whom may try to mislead consumers and regulators. Sometimes, dangers are common knowledge, although harmful products continue to be used (e.g. every year people start smoking despite knowing that it is harmful). Other problems are more subtle: investment in expensive high-tech medicine has an opportunity cost that may reduce investment in upstream improvements in public health that have a bigger positive impact.

3.2. Environmental degradation

Innovations can make dramatic contributions to environmental degradation. We are currently living in a period of accelerated environmental change, the collapse of some ecosystems and regular mass extinctions, involving the melting of Arctic and Antarctic ice, thawing of permafrost, destruction of coral reefs, irreversible fragmentation of rainforests, the expansion of deserts, increasing frequencies of droughts, floods and forest fires, and the collapse of insect and vertebrate populations (Ceballos, Ehrlich, and Dirzo 2017; Lenton et al. 2019; Kunin 2019; Díaz et al. 2019). The relationship between innovation and the environment may have both a local and a global dimension, and Hansen and Nygaard (this issue) consider a 'local issue' within a more global perspective.

3.3. Harm to society

Innovations can also harm the fabric of society. Advances in technology can lead to the erosion of privacy alongside excessive monitoring by government and industry. Leaving decisions to a computer instead of a person does not remove human biases, as programmers can build in their own biases and create new ones. Advances in automation mean that an increasing number of complex tasks and operations are being 'outsourced' to computers and algorithms, which have hidden biases, and with humans losing their capabilities as well as their overview of the process (Gardner and Bryson, this issue).

Artificial intelligence algorithms are particularly problematic if they are trained on existing datasets that reflect biases in society. In doing so, they can perpetuate and amplify racism and sexism (Didier et al. 2015).[2]

Social media also affects our self-perceptions and social interactions (Baccarella et al. 2018). It can threaten democracy through micro-targeted fake news during elections, and provides a way for nations to interfere and disrupt other nations politics. The 'click-bait' driven business model favours polarising, anger-inducing content, rather than the respectful, balanced consideration required for democratic deliberation. Social media allows populist politicians to speak directly to the population and claim to speak for them. Innovation can also harm society by increasing inequality (Witt 1996; see also Fougère and Meriläinen, this issue).

Biggi and Giuliani (this issue) observe that much research into noxious innovation has focused on work-related consequences. On the one hand, innovation may bring higher unemployment (according to fears that robots are stealing jobs). On the other hand, innovation may lead to deterioration of working conditions (e.g. Amazon warehouse workers who wear a wristband that monitors their every move; innovation in office toilets designed to become uncomfortable after sitting there for too long (Morrison 2019)). The pervasiveness of computers and cloud technologies erodes the boundaries between home spaces and work spaces, thus becoming a problem for work-life balance (Gardner and Bryson, this issue).

3.4. Harm to the economy

Innovation can also harm the economy if innovation policy is imperfectly calibrated. Excessive patent protection or monopoly power from previous innovations (e.g. from first-mover advantages, secrecy, or network effects) can harm economic dynamism and consumer welfare, and boost inequality.

For example, when the private sector succeeds in excluding others from relevant parts of the knowledge pool, or makes it difficult to understand what is patentable, it may influence the rate and direction of innovative efforts. This seems to be the case with some areas of biomedical research, where fragmented and overlapped property rights have favoured less problematic research lines in terms of intellectual property, rather than more promising ones in terms of potential impact (Heller and Eisenberg 1998).

4. Unanswered questions

Three unanswered questions can be mentioned here. First, can government ever apply the precautionary principle (Stirling 2017) fast enough to protect against harmful side-effects on e.g. public health? Innovative substances and technologies are increasing artificial and complex, and their integration into society and the environment is often not well understood. Given that innovation is fundamentally uncertain and involves experimentation, there may be gaps in the regulation of moving from scientific discoveries to exploratory applications in the real world (e.g. applying exploratory new medical techniques on real-world patients in unethical ways; see Saemundsson and McKelvey, this issue).

Second, if innovation gets us into this mess, can we expect innovation to get us out of it? Innovation may give us the virus, as well as the vaccine (although many may die in between). Or does innovation only lead us into ever increasing difficulties (e.g. road traffic no longer means horse manure on our streets, but exhaust fumes instead). We agree, for example, with the need to give subsidies to renewable energy R&D to reduce the harm from fossil fuels (Acemoglu et al. 2012), but alternatives that are, by almost any definition, improvements upon fossil fuels, are nonetheless bringing new problems: for example, wind turbine blades are piling up in landfills (Martin 2020), and there are problems of solar e-waste (Hansen and Nygaard, this issue).

Third, accidents happen. What is the acceptable rate of 'normal accidents' such as Chernobyl? If it is above zero, innovation will be accompanied by catastrophes that are unpredictable and increasingly large (even if the setting-up of preventative procedures could sometimes help to avoid some catastrophes). If it is zero, then this 'risk-averse' stance could put an absolute halt on some types of innovation.

5. Overview of the papers

The papers in this study (see Table 1) are a mix of quantitative analysis (Biggi and Giuliani), case studies at the level of industries (Hansen and Nygaard) or specific persons

[2]Software engineers are rarely trained to deal with these problems. It is therefore concerning that they were surprised that a chat-bot trained on social media comments would end up a feminist-hating, Hitler-supporting bigot within 24 hours, when such an outcome was obvious to others.

Table 1. Papers in this special issue.

Authors	Title	Method	Theme
Biggi, G., Giuliani E.	The Noxious Consequences of Innovation: What do we know?	Literature review Bibliometric analysis	State of the art of the literature on the dark side of innovation
Hansen, U., Nygaard, I.	The dark side of the sun: environmental upgrading in the off-grid solar PV value chain.	Case study	A local and global perspective on new technologies and the environment
Gardner, E., Bryson, J.	The Dark Side of the Industrialisation of Accountancy.	60 semi-structured interviews with UK accountancy firms	How digital innovations are disrupting the accountancy profession, bringing new problems.
Saemundsson, R., McKelvey, M.	Negative Unintended Consequences as a Counterbalance to Innovation.	Case study	The regulation of medical innovation to balance between novelty and risk
Fougère, M., Meriläinen, E.	Exposing Three Dark Sides of Social Innovation through Critical Perspectives on Resilience.	Theoretical discussion	Social innovation and resilience discourses may lead to unwanted decisions/outcomes in real situations

and events (Saemundsson and McKelvey), interviews (Gardner and Bryson) and theoretical reasoning (Fougère and Meriläinen).

Biggi and Giuliani begin with a review of literature that acknowledges the potentially harmful side of innovation. They find 125 articles in 89 journals. This suggests that the evidence base is scattered, and perhaps provides some justification for the need to publish a dedicated special issue on the topic, to take stock of the dispersed literature. Biggi and Giuliani (2020)'s analysis yields 5 clusters of 'dark side of innovation' papers. They find that most of the papers in their literature review are in the clusters 'Work-related consequences of technology acceptance' and 'Unsustainable transitions.' It is therefore interesting to observe that 'work-related consequences of technology acceptance' are also discussed in our Special Issue (in particular, by Gardner and Bryson (2020)).

Meanwhile our Special Issue also makes a contribution to research that could be categorised as 'emerging technologies', Saemundsson and McKelvey (2020) focusing on innovation in clinical practices, with an eye to their global implications, as in the contribution of Hansen and Nygaard (2020) focusing on solar panel technologies.

Fougère and Meriläinen (2020) round off the Special Issue by discussing the important themes of social innovation and the resilience of marginalised communities. The priority given to the economic measurement of the impact of societal innovation in the contemporary policy framework may emphasise their positive side, hiding the possible negative effects on specific groups. The authors warn about the risk that the social innovation and resilience discourses may be hijacked by powerful actors.

The dark side of innovation is an underexplored field of research that offers a wide range of opportunities. With this special issue we have tried to put together a set of results and perspectives to conceptualise the downsides of innovation. We hope this will favour research in the field, as we seek a more balanced understanding to innovation.

Acknowledgments

We are grateful to the Editor, Bram Timmermans, for many helpful comments. Any remaining errors are ours alone.

Disclosure statement

No potential conflict of interest was reported by the authors.

References

Acemoglu, D., P. Aghion, L. Bursztyn, and D. Hemous. 2012. "The Environment and Directed Technical Change." *American Economic Review* 102 (1): 131–166. doi:10.1257/aer.102.1.131.

Akimoto, H. 2003. "Global Air Quality and Pollution." *Science* 302 (5651): 1716–1719. doi:10.1126/science.1092666.

Baccarella, C. V., T. F. Wagner, J. H. Kietzmann, and I. P. McCarthy. 2018. "Social Media? It's Serious! Understanding the Dark Side of Social Media." *European Management Journal* 36 (4): 431–438. doi:10.1016/j.emj.2018.07.002.

Bausinger, T., E. Bonnaire, and J. Preuß. 2007. "Exposure Assessment of a Burning Ground for Chemical Ammunition on the Great War Battlefields of Verdun." *Science of the Total Environment* 382 (2–3): 259–271. doi:10.1016/j.scitotenv.2007.04.029.

Biggi, G., and E. Giuliani. 2020. *The Noxious Consequences of Innovation: What Do We Know?* Industry & Innovation, this issue.

Ceballos, G., P. R. Ehrlich, and R. Dirzo. 2017. "Biological Annihilation via the Ongoing Sixth Mass Extinction Signaled by Vertebrate Population Losses and Declines." *Proceedings of the National Academy of Sciences* 114 (30): E6089–E6096. doi:10.1073/pnas.1704949114.

Conca, J. 2020. "Blood Batteries - Cobalt and the Congo." [online] *Forbes*. https://www.forbes.com/sites/jamesconca/2018/09/26/blood-batteries-cobalt-and-the-congo/#2a6d5559cc6e

Cooper, P. 2018 25 April. https://twitter.com/PaulMMCooper/status/989100350044082176

Cotti, C., E. Nesson, and N. Tefft. 2018. "The Relationship between Cigarettes and Electronic Cigarettes: Evidence from Household Panel Data." *Journal of Health Economics* 61: 205–219. doi:10.1016/j.jhealeco.2018.08.001.

Cressey, D. 2016. "The Plastic Ocean." *Nature* 536 (7616): 263–265. doi:10.1038/536263a.

Díaz, S., J. Settele, E. S. Brondízio, H. T. Ngo, J. Agard, A. Arneth, P. Balvanera, et al. 2019. "Pervasive Human-driven Decline of Life on Earth Points to the Need for Transformative Change." *Science* 366: 6471. doi:10.1126/science.aax3100.

Didier, C., W. Duan, J. P. Dupuy, D. H. Guston, Y. Liu, J. A. L. Cerezo, D. Michelfelder, et al. 2015. "Acknowledging AI's Dark Side." *Science* 349 (6252): 1064–1065. doi:10.1126/science.349.6252.1064-c.

Ebbeling, C. B., D. B. Pawlak, and D. S. Ludwig. 2002. "Childhood Obesity: Public-health Crisis, Common Sense Cure." *The Lancet* 360 (9331): 473–482. doi:10.1016/S0140-6736(02)09678-2.

Fougère, M., and E. Meriläinen. 2020. *Exposing Three Dark Sides of Social Innovation through Critical Perspectives on Resilience*. Industry & Innovation, this issue.

Franklin-Wallis, O. 2019. "'Plastic Recycling is a Myth': What Really Happens to Your Rubbish?" [online] *The Guardian*. https://www.theguardian.com/environment/2019/aug/17/plastic-recycling-myth-what-really-happens-your-rubbish

Gardner, E., and J. Bryson. 2020. *The Dark Side of the Industrialisation of Accountancy*. Industry & Innovation. Forthcoming, this issue.

Hansen, U., and I. Nygaard. 2020. *The Dark Side of the Sun: Solar E-waste and Environmental Upgrading in the Off-grid Solar PV Value Chain*. Industry & Innovation, submitted.

Heller, M. A., and R. S. Eisenberg. 1998. "Can Patents Deter Innovation? The Anticommons in Biomedical Research." *Science* 280 (5364): 698–701. doi:10.1126/science.280.5364.698.

Holbreich, M., J. Genuneit, J. Weber, C. Braun-Fahrländer, M. Waser, and E. Von Mutius. 2012. "Amish Children Living in Northern Indiana Have a Very Low Prevalence of Allergic Sensitization." *Journal of Allergy and Clinical Immunology* 129 (6): 1671–1673. doi:10.1016/j.jaci.2012.03.016.

Jasanoff, S. 2016. *The Ethics of Invention: Technology and the Human Future*. WW Norton & Company: London and New York.

Kunin, W. E. 2019. "Robust Evidence of Declines in Insect Abundance and Biodiversity." *Nature* 574: 641–642. doi:10.1038/d41586-019-03241-9.

Lee, S., E. P. Hess, C. Lohse, W. Gilani, A. M. Chamberlain, and R. L. Campbell. 2017. "Trends, Characteristics, and Incidence of Anaphylaxis in 2001–2010: A Population-based Study." *Journal of Allergy and Clinical Immunology* 139 (1): 182–188. doi:10.1016/j.jaci.2016.04.029.

Lenton, T. M., J. Rockström, O. Gaffney, S. Rahmstorf, K. Richardson, W. Steffen, and H. J. Schellnhuber. 2019. "Climate Tipping Points—too Risky to Bet Against." *Nature* 575: 592–595. doi:10.1038/d41586-019-03595-0.

Levine H, Jørgensen N, Martino-Andrade A, Mendiola J, Weksler-Derri D, Mindlis I, Pinotti R, Swan SH. 2017. "Temporal Trends in Sperm Count: A Systematic Review and Meta-regression Analysis." *Human Reproduction Update* 23 (6): 646–659.

Lim, X. 2019. "Tainted Water: The Scientists Tracing Thousands of Fluorinated Chemicals in Our Environment." *Nature* 566 (7742): 26–30. doi:10.1038/d41586-019-00441-1.

Luzzati, T., A. Parenti, and T. Rughi. 2018. "Economic Growth and Cancer Incidence." *Ecological Economics* 146: 381–396. doi:10.1016/j.ecolecon.2017.11.031.

Martin, C. 2020. "Wind Turbine Blades Can't Be Recycles, so They're Piling up in Landfills?" [online] *Bloomberg Green*. //www.bloomberg.com/news/features/2020-02-05/wind-turbine-blades-can-t-be-recycled-so-they-re-piling-up-in-landfills

Morrison, R. 2019. "Say Goodbye to Comfort Breaks! New Downward-tilting Toilets are Designed to Become Unbearable to Sit on after Five Minutes." [online] *Daily Mail*. https://www.dailymail.co.uk/sciencetech/article-7801245/Toilets-tilted-downwards-13-degrees-stop-workers-spending-long-loo.html

Nelson, R. R. 1962. *The Rate and Direction of Inventive Activity*. NBER, Princeton University Press. Princeton, New Jersey.

Nelson, R. R. 1977. *The Moon and the Ghetto*. New York: Norton.

Nobel, J. 2020. "America's Radioactive Secret." [online] *RollingStone*. https://www.rollingstone.com/politics/politics-features/oil-gas-fracking-radioactive-investigation-937389/

Proctor, R. N. 2012. "The History of the Discovery of the Cigarette–lung Cancer Link: Evidentiary Traditions, Corporate Denial, Global Toll." *Tobacco Control* 21 (2): 87–91. doi:10.1136/tobaccocontrol-2011-050338.

Rothenberg, J. 1970. "The Economics of Congestion and Pollution: An Integrated View." *American Economic Review* 60 (2): 114–121.

Saemundsson, R., and M. McKelvey. 2020. *Negative Unintended Consequences as a Counterbalance to Innovation*. Industry and Innovation, forthcoming.

Soffritti, M., F. Belpoggi, E. Tibaldi, D. D. Esposti, and M. Lauriola. 2007. "Life-span Exposure to Low Doses of Aspartame Beginning during Prenatal Life Increases Cancer Effects in Rats." *Environmental Health Perspectives* 115 (9): 1293–1297. doi:10.1289/ehp.10271.

Stilgoe, J. 2020. *Who's Driving Innovation? New Technologies and the Collaborative State*. Palgrave. Palgrave / Springer Nature Switzerland: Cham, Switzerland.

Stirling, A. 2017. "Precaution in the Governance of Technology." In *The Oxford Handbook of Law, Regulation, and Technology*, edited by R. Brownsword, E. Scotford, and K. Yeung, p. 645-669. Oxford, UK: Oxford University Press. doi:10.1093/oxfordhb/9780199680832.001.0001.

Svensson, S., J. L. Richter, E. Maitre-Ekern, T. Pihlajarinne, A. Maigret, and C. Dalhammar. 2018. "The Emerging 'Right to Repair' Legislation in the EU and the U.S." Paper presented at Going Green CARE INNOVATION 2018, Vienna, Austria.

Tenner, E. (1997). Why things bite back: Technology and the revenge of unintended consequences. New York: Alfred A Knopf.

Warner, N. R., C. A. Christie, R. B. Jackson, and A. Vengosh. 2013. "Impacts of Shale Gas Wastewater Disposal on Water Quality in Western Pennsylvania." *Environmental Science & Technology* 47 (20): 11849–11857. doi:10.1021/es402165b.

Witt, U. 1996. "Innovations, Externalities and the Problem of Economic Progress." *Public Choice* 89 (1–2): 113–130. doi:10.1007/BF00114282.

Woerner, A. 2015. "What are Natural Flavors, Really?" [online] *CNN Health*. https://edition.cnn.com/2015/01/14/health/feat-natural-flavors-explained/index.html

Zingales, L. 2015. "Presidential Address: Does Finance Benefit Society?" *Journal of Finance* 70 (4): 1327–1363. doi:10.1111/jofi.12295.

The noxious consequences of innovation: what do we know?

Gianluca Biggi and Elisa Giuliani

ABSTRACT

In spite of being considered an undisputed engine of growth, innovation can have noxious consequences for society and the environment. Using bibliometric techniques (i.e. bibliographic coupling and co-citation analysis), we conduct a review of the extant research on the noxious impacts of innovation. Although this is a relatively recent field of enquiry, we identified five strands of scholarly research, which, based on their focus, we have labelled: (A) *Work-related consequences of technology acceptance;* (B) *Unsustainable transitions;* (C) *Innovation and growth downside effects;* (D) *The dangers of emerging technologies* and (E) *Open innovation's dark side.* We discuss the core ideas and research agendas in these research strands and the intellectual antecedents of each sub-community, and conclude by suggesting avenues for future research.

1. Introduction

Most of the scientific work in the field of innovation studies starts from the assumption that innovation contributes to firm growth and survival and to the growth of the economic system more broadly. Indeed, the perennial Schumpeterian gale of creative destruction (Schumpeter 1942) has underpinned decades of research on how to promote innovation to boost economic development processes. The positive relationship between science, technology & innovation and economic growth has been so widely acknowledged both in the field of evolutionary economics (Dosi et al. 1988), and in modern theories of endogenous growth (e.g. Romer 1986; Aghion and Howitt 1992, 1998), that it is currently taken for granted. Innovation is seen as a key engine of competitiveness at the firm, industry and national levels (e.g. Freeman 1988; Lundvall 1992). Accordingly, most of the existing research emphasises the 'bright side' of technological change and overlooks its social and environmental downsides (Dosi 2013).

It is true that the political agenda has been oriented towards stimulating economic competition, wealth creation and productivity through technological progress, which, in turn, have improved some of the economic and social indicators in most market economies. However, at the same time, innovation-induced industrial activities have had important negative consequences, in terms of both environmental disasters and deterioration of human rights (Giuliani 2018). Indeed, modern economic growth is based on intensive use of non-renewable energy sources, industrial chemicals and the production of massive non-recyclable waste. For instance, Luzzati, Parenti, and Rughi (2018) show that increased incidence of cancers is becoming a public health issue in most of advanced economies, allegedly due to their population's exposure to the toxic emissions of industrial activities. Use of chemical inputs, such as pesticides, are another example of failed 'economic progress' as a result of technological change; it strikes that the traces of 12 of the most toxic chemicals ever invented, which were banned after the 2001 Stockholm Convention for their noxious impacts on human health and the environment, can still be found in the most remote areas of the world, such as deserts, open ocean and arctic regions, where there is no human activity or chemicals applications (Lohmann et al. 2007). Innovations have also had perilous consequences in finance, for their calamitous contributions to the 2008 financial crisis (Dosi 2011, 2013): financial innovations such as sub-prime mortgages, collateralised debt obligations (CDO) and credit default swaps (CDS) all played a central role in creating the crisis, giving rise to a process of 'destructive creation' (Soete 2013). In another context, scholars have noted that, at times, technological change has also had detrimental effects on employment which have not been entirely compensated by the wave of opportunities coming from the change. For instance, Korinek and Stiglitz (2017) recently showed that the diffusion of artificial intelligence and other forms of worker-replacing technological progress, could have a negative impact on both income distribution and employment.

It has thus become progressively clear that innovation has had unintended direct effects and negative side effects (Merton 1936) on society and on the environment. That is, effects contrary to what was originally intended or noxious impacts alongside the desired effects. As a result, researchers are calling for more 'Responsible Innovation' (RI) (Hellström 2003; Owen, Macnaghten, and Stilgoe 2012), understood as innovation that takes care 'of the future through collective stewardship of science and innovation in the present' (Stilgoe, Owen, and Macnaghten 2013, 1570), which is introducing a new international policy agenda that includes the European Commission Horizon 2020 research and innovation programme – Responsible Research and Innovation (RRI) – among other initiatives.

More generally, innovation scholars are giving prominence to social and environmental issues, in their analyses and policy discourse (Schot and Steinmueller 2018). For instance, research on transformative change and sustainability transitions (Geels 2004; Meadowcroft 2011; Markard, Raven, and Truffer 2012) is including harnessing STI to social and environmental issues and the pursuit of Sustainable Development Goals (SDGs) (United Nations 2015) such as protecting the earth's system and improving human living conditions. Despite supranational frameworks, such as the United Nations Global Compact (United Nations 2014) and the World Business Council for Sustainable Development (WBCSD), the pace of climate change and other negative impacts on the environment and society, are increasing (Griggs et al. 2013), with some of the boundaries calculated by scholars as necessary for the earth system to be sustainable, already crossed (Steffen, Crutzen, and McNeill 2007; Rockström et al. 2009).

Over recent years, the socio-environmental challenges of innovation have been attracting the attention of researchers in a range of scholarly traditions; however, few efforts, to our knowledge, have been aimed at making sense of the existing literature – and especially work on the noxious impacts of innovation activities. To address this gap, we provide a systematic review of the literature on this salient topic. In particular, we aim to identify (a) the dominant research themes and (b) their intellectual precursors. Our review covers 125 empirical contributions, published in economics, business and management-related journals, between 1991[1] and the end of 2017. Our analysis uses bibliometric techniques for science mapping, to reduce the reviewer subjectivity bias typical of qualitative reviews (Vogel and Güttel 2013). Bibliographic coupling and co-citation analysis allow a visualisation of the network of publications grouped in clusters of thematically related works.

The remainder of the paper is structured as follows: Section 2 presents the methodology employed for the bibliographic analysis. Section 3 presents the main results of the analysis and Section 4 concludes and makes some suggestions for future research directions.

2. Literature review: data and methods

We conducted a systematic literature search, aimed at a high level of inclusiveness, and applied bibliometric analyses to provide a full, thematic overview of the existing knowledge on the noxious effects of innovation. Bibliographic records, including keywords and citations, are useful elements of scientific work (Garfield, Malin, and Small 1983) and, recently, have been acknowledged as reliable methods to study the scope and dynamics of scientific fields in innovation studies (e.g. Schildt, Zahra, and Sillanpää 2006; Raasch et al. 2013; van der Have and Rubalcaba 2016; van Oorschot, Hofman, and Halman 2018). In contrast to other techniques, bibliometrics facilitates more objective and reliable analyses and provides a 'big picture' of extant research (Crane 1972) and science mapping. One of its advantages is that it reveals the structural and dynamic aspects of scientific research (Noyons, Moed, and Luwel 1999; Börner, Chen, and Boyack 2003). A systematic review approach differs from traditional narrative reviews by adopting a replicable, scientific and transparent protocol, aimed at minimising bias through an exhaustive literature search (Tranfield, Denyer, and Smart 2003).

2.1. *Article search*

We constructed a search query using the WoS database. We decided to use the Social Science Citation Index of Clarivate Analytics Web of Science (WoS SSCI) because its 'carefully selected and evaluated collection delivers to users the most influential scientific research information from the twentieth century'.[2] We used search terms such as harm* OR unpredictable OR negative OR hazard* AND innovat* OR technolog* and several combinations of related and similar terms (see Appendix A for the full list). Since we are interested in the literature focused on the noxious effects of innovation on the functioning of economic systems and their actors, we searched in the management, business and economics literatures, which identified 328 documents. Following Thyer (2008), we selected only peer-reviewed academic journal articles. We excluded conference papers, book chapters and books based on the fact that peer-reviewed academic journal articles are normally considered to be at the frontier of knowledge and, compared to these other

[1]The publication date of the oldest article in our dataset.
[2]Source: http://wokinfo.com/products_tools/multidisciplinary/webofscience/ssci/.

types of publications, apply the most advanced methodological standards. Next, to ensure the relevance of each article for our purposes, the abstracts, keywords and introductions were checked manually by one of the authors. This allowed identification and elimination of false-positive articles (i.e. articles not related to our topic of interest). We performed a similar, additional search in Scopus. After manually comparing all the search results, using the criteria described above, we included additional 14 records. To ensure study reliability, the full texts of the selected articles were coded manually by one of the authors. The overall search process resulted in a sample of 125 bibliographic records (hereinafter 'bibliographic sample'), covering more than 6,800 cited references and 950 unique keywords (see Figure 1).

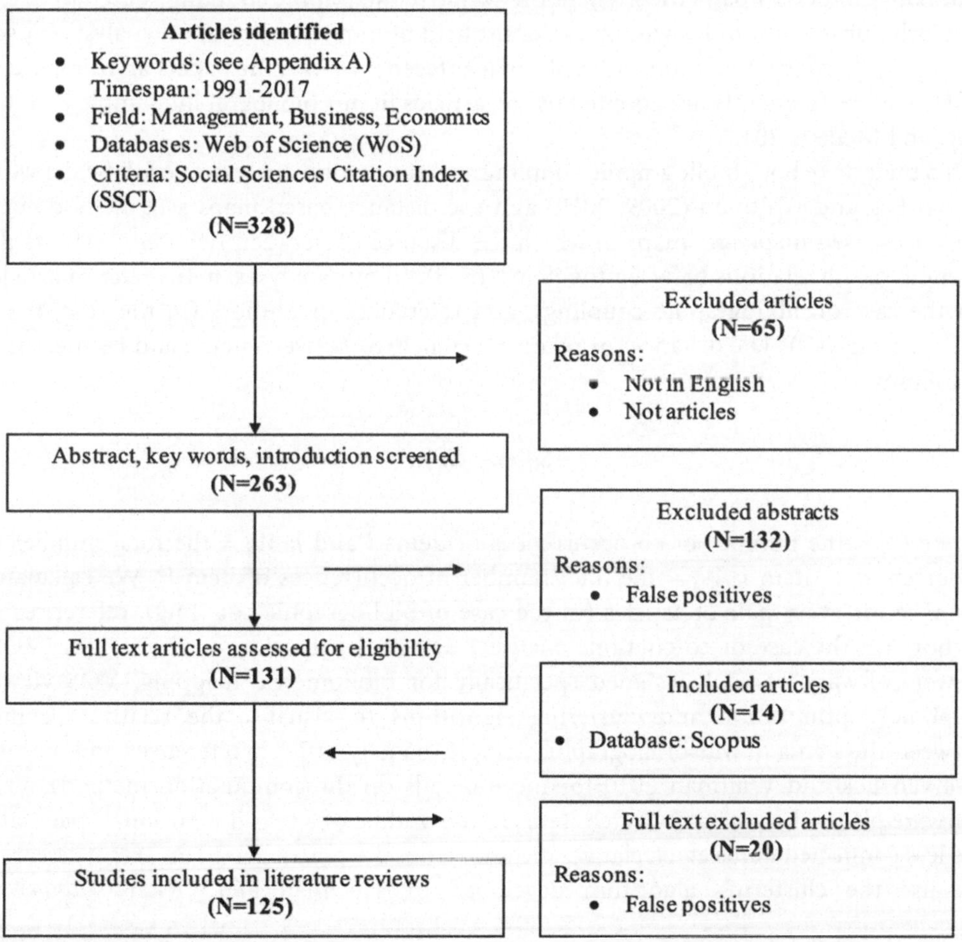

Figure 1. Data-collection process.

Source: Authors' own elaboration.

2.2. Methods: bibliographic coupling and co-citation analysis

To perform our literature review, we apply two increasingly popular bibliometric methods: bibliographic coupling and co-citation analysis (see, e.g. van der Have and Rubalcaba 2016; van Oorschot, Hofman, and Halman 2018). The former clusters articles from a bibliographic sample based on their common references (i.e. references appearing in the reference list of more than 1 article in the bibliographic sample; hereinafter 'common citations') (Boyack and Klavans 2010). The more two articles share common citations, the higher is their 'coupling strength', which means that they build on a mutual knowledge base, and can thus be considered as being the expression of the same scholarly community (on this, see: Kessler 1963; Martyn 1964). Based on this methodology, we seek to identify the sub-communities of scholars investigating the noxious impacts of innovation.

The latter – co-citation analysis – does not cluster the bibliometric sample's articles (as in the case of bibliographic coupling), but these articles' references and these references' authors (for an application see also: Boyack and Klavans 2010; Meyer et al. 2014; Kovacs, Van Looy, and Cassiman 2015). Therefore, while bibliographic coupling is used to detect scholarly sub-communities within a relevant field of inquiry, co-citation analysis is used to identify its conceptual roots or intellectual antecedents – i.e. the papers and/or authors that are more frequently are co-cited by the articles in our bibliographic sample (Youtie, Kay, and Melkers 2013).

To undertake both bibliographic coupling and co-citation analysis, we follow the work of van Eck and Waltman (2009, 2010) and use distance-based maps as a methodology. Distance-based maps are 'maps in which the distance of between two items reflects the strength of the relations between the items' (525). In our analyses, items refer to articles (in the case of bibliographic coupling), and references or authors (in the case of co-citations analysis). The distance (or relative distance) S_{ij} between item i and j is measured as follows:

$$S_{\mathrm{ij}} = \frac{C_{ij}}{W_i W_j}$$

where C_{ij} = the number of co-occurrences of items i and j; W_i = the total number of occurrences of item i; W_j = the total number of occurrences of item j.[3] We calculated distance for every pair of articles (in the case of bibliographic coupling), references or authors (in the case of co-citations analysis) using Visualisation of Similarities (VOS) viewer software,[4] a tool designed specifically for bibliometric mapping. VOS viewer combines optimisation and clustering algorithms to visualise the relative distance between the items of our bibliographic sample, namely articles, references and authors (see van Eck and Waltman 2010 for more details on the computation method). VOS software places the most connected items in the middle of a two-dimensional space, and the less connected items at a (relative) distance from the centre. Based on extant research, we use the clustering algorithm developed by Newman and Givan's (2004) as

[3]In the case of bibliometric coupling, co-occurrence (C_{ij}) measures the number of citations that articles i and j have in common. In the case of co-citation analysis co-occurrence (C_{ij}) refers to the number of times that articles i and j both cite a given reference or author.

[4]Source: http://www.vosviewer.com.

implemented in VOS software (Cobo et al. 2011). Such algorithm can be parametrised in VOS viewer by a resolution parameter that is able to detect also small size clusters, which is ideal in our context of a newly emerging field of inquiry. A higher resolution parameter implies an increment in the number of potentially detected clusters (Waltman and van Eck 2013). We followed a conservative approach by using the default parameter for co-citation analysis, while we adjusted the parameter for the bibliographic coupling to 0.75 (against the default 1.0 value provided by VOS) because after a qualitative assessment we considered this latter value to yield the most meaningful clusters and provide a clearer and more precise overview of the existing literature.

Our analyses allow us to identify clusters which are highlighted on a map using different depths of shading. The more proximate the items on the map, the more the reader will be able to identify either sub-communities (via bibliographic coupling) or meaningful conceptual roots (via co-citation analysis). When we performed bibliographic coupling, we apply no restrictions on minimum number of common citations. From the initial sample of 125 articles, 107 (i.e. 85,6% of the original dataset) were grouped in clusters which we labelled according to their main research theme to facilitate the interpretation.[5] Finally, in the case of the co-citation analysis, since our bibliographic sample includes 6,869 cited references and 5,453 cited authors, we have decided to restrict our focus to references or authors co-cited more than eight times within the sample, in order to select out the least influential ones.[6]

3. Findings

Our literature review includes 125 articles published in 89 journals during the period 1991–2017; the top (highest number of published papers on the topic) 7 journals (namely, *Technovation, Industrial and Corporate Change, Journal of Management Information System, R&D Management, Research Policy, Technological Forecasting and Social Change, Technology Analysis & Strategic Management)* account for around 30% of the articles. This suggests that research interest on this subject is more prevalent in innovation-related journals rather than more general mainstream management and economics outlets, but also that there is a variety of fields and intellectual traditions that feed into this subject, which has yet to emerge as dominant knowledge related to one or a few specific journals and their communities. This clearly makes this Special Issue contribution particularly timely, since it documents, for the first time, evidence of the mounting interest among scholars on this topic. Figure 2 depicts the distribution over time of the articles included in our review. The observation period can be split into a first relatively long period of early emergence (1991–2005) when we found only a limited number of publications on the effects of harmful innovation (i.e. no more than 3 papers per year), and a period of 'take-off', starting in 2006 and peaking after 2011, although with some fluctuations until the last two years of our analysis, when research on harmful innovation became much more prominent.

---

[5]VOS viewer dropped 18 articles because none of their references had at least one 'common citation' with the other articles of the bibliographic sample. They were isolated nodes in the bibliographic coupling network, which suggests that they were not part of any of the emerging sub-communities identified by the analysis.

[6]To define cut-off points we followed van Eck and Waltman (2010). As a robustness check, we considered alternative cut-off points (i.e. 6, 7 and 9) with understandably little variations in the results.

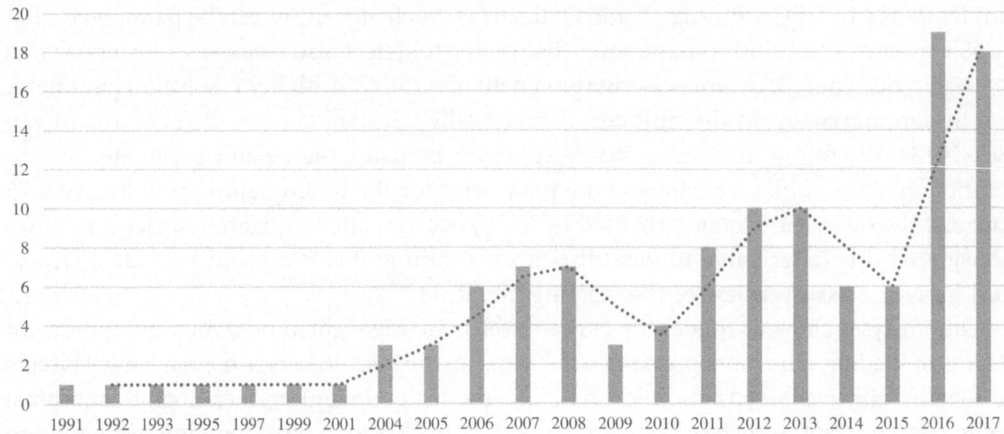

Figure 2. Publications per year 1991–2017.
Source: Authors' own elaboration.

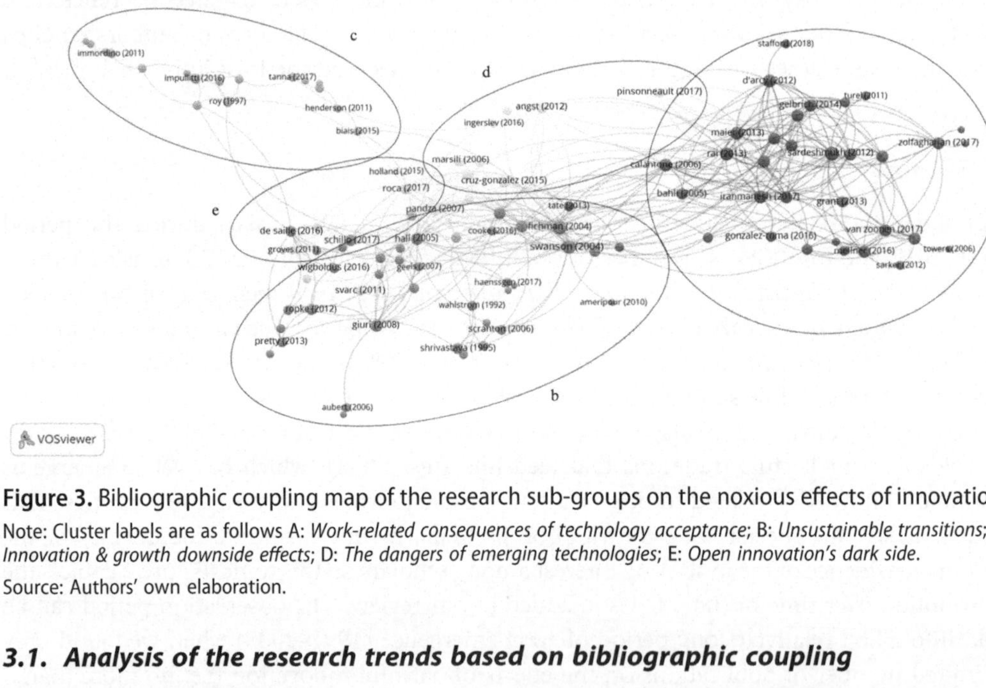

Figure 3. Bibliographic coupling map of the research sub-groups on the noxious effects of innovation.
Note: Cluster labels are as follows A: *Work-related consequences of technology acceptance*; B: *Unsustainable transitions*; C: *Innovation & growth downside effects*; D: *The dangers of emerging technologies*; E: *Open innovation's dark side*.
Source: Authors' own elaboration.

3.1. Analysis of the research trends based on bibliographic coupling

The results of bibliographic coupling of the articles included in our dataset are depicted in Figure 3. We identified a modular network, characterised by five distinct, but inter-related clusters. We labelled these clusters based on their main research theme, as: *Work-related consequences of technology acceptance* (cluster A); *Unsustainable transitions* (cluster B); *Innovation & growth downside effects* (cluster C); *The dangers of emerging technologies* (cluster D); *Open innovation's dark side* (cluster E). Our publications network (Figure 3), consists of vertices that represent the bibliographic-paired articles. The distances between vertices correspond to the likelihood of bibliographic coupling (i.e. the closer the vertices in the network, the higher the common citations). We summarise the salient results and issues of inquiry in each sub-community in the sections that follow.

3.1.1. Cluster A: work-related consequences of technology acceptance

Cluster A is in a central position in the bibliographic coupling network in Figure 3. Research included in this cluster studies the adverse consequences of the adoption of Information and Communication Technologies (ICT) in organisations - with a particular focus on pervasive ICT technologies (i.e. technologies that can be used almost anytime and anywhere). The adverse consequences considered relate, mainly, to psychological effects on employees after technology adoption, related to job satisfaction and turnover intention (Ragu-Nathan et al. 2008), work-life conflicts (Golden 2006) and burn-out (Maslach, Schaufeli, and Leiter 2001).

For instance, Turel, Serenko, and Giles (2011) provide an empirical analysis of North American firms and show how addiction to workplace technologies (e.g. mobile technologies; email) can decrease productivity and organisational commitment and increase turnover intention and work–family conflicts. The addiction symptoms identified refer to employees' inability to reduce their use of workplace technologies due to psychological dependency. In this context, Turel et al. show that these employee additions threaten their social lives (family and friends) and cause self-neglect. Golden, Veiga, and Dino (2008) find similar results for the impacts of remote working technologies, among large high-tech companies that employ remote workers permanently. The authors identify a measure of professional isolation to investigate its connection to job performance and turnover intentions. The more recent studies in this cluster focus on the impacts of work-related social media use on employees. For instance, van Zoonen, Verhoeven, and Vliegenthart (2016) argue that work-related social media use can trigger role pressures that have negative psychological effects leading to emotional exhaustion among employees.

3.1.2. Cluster B: unsustainable transitions

Cluster B displays significant growth in the number of publications per year, compared to the other clusters. Most of the articles of this cluster have a meso-level unit of analysis and examine socio-technical systems (e.g. electricity, agro-food, transportation, etc.) These works address the unintended and harmful effects of innovation and investigate the complex interactions between the biophysical, social, economic and institutional factors resulting from the introduction of an innovation in the current socio-technical system. These studies tend to emphasise the directionality of transitions, arguing that innovations can have a tendency to develop in unsustainable ways. For instance, in an ethnographic study of Danish households, Røpke, Christensen, and Jensen (2010) explore unsustainable patterns of consumption of ICT. The study shows that the energy that is consumed by these technologies (including the most energy efficient ones) keeps increasing with their incorporation in everyday social practices. As a result, the energy reductions derived from more efficient technologies are outweighed by the energy consumption of ICT in households. In a study of the agricultural system, Wigboldus et al. (2016) investigate the scaling of agricultural-improving technologies such as disease-resistant seeds, permaculture cultivation practices and automated milking systems. They show how the scaling up of these technologies may pose a series of undesirable effects such as unanticipated environmental degradation effects, poor farm worker labour conditions and loss of control by farming communities over seeds. For example, pesticides give positive economic returns in the short run, but generate negative externalities in the form of environmental degradation and adverse impacts on human health. The adoption of these detrimental innovations is ascribed (mostly) to the fact that the agricultural system became locked into a path-dependent production model which has developed along a direction that it is difficult to change despite its being environmentally and socially unsustainable.

Of particular interest is that the research in this cluster, stresses that, in some cases, even innovations designed to contribute positively to the environment – see, e.g. improvements to materials and improved energy efficiency in buildings – may cause significant unintended negative impacts on society, including greater consumption and increased waste and pollution, giving rise to the so-called 'rebound (or take-back) effect' (Frondel 2004; Herring 2004). Herring and Roy (2007) study is a fitting example of this. They examine the rebound effect in an empirical study focused on the factors that influence consumers' adoption of low and zero carbon technologies. They show that adopting energy-efficient technologies can result in reduced energy consumption, in the short term, of between 10% and 20%; however, this positive effect is soon outweighed by rising levels of consumption. Rebound effects are observed, also, in the transportation system, as noted by Wang, Zhou, and Zhou (2012), who investigate the direct rebound effect for passenger transport in China in the period 2002–2007 and show that energy-saving transport systems reduce transport costs, but that the energy reductions are offset by increased demand for public transport due to its lower costs.

3.1.3. Cluster C: innovation & growth downside effects

Cluster C mostly includes economics research on the consequences of economic growth fuelled by innovation. The articles in this cluster tend to address the negative side-effects of innovation from a macro-level perspective. Several such studies show the impact of technological change on the economic, social and environmental spheres (see Roy 1997; Dilaver 2014; Klingelhöfer 2017). For instance, Managi and Kumar (2009) estimate trade-induced technological changes in 76 countries, over the period 1963–2000, and show that such changes, which constitute around a third of all technological change, not only have a negative effect on GDP but also increase pollution (see also Di Maria and Smulders 2004).

Another topic covered in this cluster is the social and economic consequences of financial innovations, understood as new financial instruments (e.g. sub-prime mortgages, collateralised debt obligations and credit default swaps). Henderson and Pearson (2011), in their study of 64 popular financial instruments issued during 2001–2005, provide striking evidence that financial institutions designed innovative financial instruments that exploit investors' misunderstandings, and show that the overall returns on these instruments were less than the risk-free rate. Pérignon and Vallée (2017), in a study of French local government innovative and high-risk loan choices, provide evidence of short-term biases in public finance decisions.

Finally, work in this cluster addresses the relationship between innovation, employment and economic inequality. As noted by Breau, Kogler, and Bolton (2014) in their empirical investigation of 85 Canadian cities, innovation matters for explaining differences in patterns of wage inequality: the authors show that, compared to non-urban areas, metropolitan areas with high levels of patent counts and employment in knowledge-based industries, display a more unequal wage distribution. These results are in line, to an extent, with Lee and Rodríguez-Pose (2013) comparative study of European regions and US cities. They find that the overall level of innovation drives inequality in European regions – although not in US cities. Related to this, several scholars have investigated the relationship between technological change and employment. For instance, Jenkins (2008) examines the case of South Africa in the 1990s and finds that both technological change and trade liberalisation had a negative impact on employment.

3.1.4. Cluster D: the dangers of emerging technologies

Despite being the smallest cluster in our bibliographic coupling network, for number of publications, this cluster shows the strongest growth in their number in recent years. Most contributions in this cluster examine the scientific uncertainties surrounding emerging technologies (i.e. nanotechnologies, big data, the Internet of Things, cognitive computing, virtual technologies and social media) and the role of policy in keeping pace with these industry developments. There is a set of studies that focus specifically on the potential harmful impacts and related societal challenges of nanotechnologies. Nanotechnologies are considered to be a threat to public health and the environment, because of the impossibility to assess the risks associated to these emerging technologies. Lu et al. (2012) note that public health risks include large doses of nanoparticles that can accumulate in cells and cause toxic responses, while the environmental risks include the effect on the environment of nanostructures via bioaccumulation and creation of non-biodegradable pollutants. Along the same lines, an in-depth survey of the UK nanotechnology industry, by Groves et al. (2011), warns against the uncertainties and risks associated to nanotechnology with specific reference to the toxicological effects of nanomaterials and related products, which neither manufacturers nor governments appear to be well prepared to deal with in a systematic and timely way.

This cluster includes articles that seek to identify and assess the possible unintended negative consequences of big data, the Internet of Things and social media, by proposing frameworks to help organisations to understand the probabilities that these emerging technologies could cause harm to societal interactions and the environment (Wilburn and Wilburn 2016). For instance, using the example of the emerging technology of metal additive manufacturing, which could pose unknown health and environmental risks, Roca et al. (2017) propose a regulatory framework that adapts dynamically to the emergence of new threats posed by scientific and technological advances in the field of additive manufacturing.

3.1.5. Cluster E: open innovation's dark side

Cluster E examines the negative consequences of open innovation as a powerful search strategy, encompassing the generation, capture and employment of new knowledge at firm-level (West and Bogers 2014; Chesbrough, Vanhaverbeke, and West 2006). Whereas the negative impacts observed in the other clusters affect predominantly individuals and are social and environmental hazard related, this cluster focuses on the negative impacts of organisational innovation on the performance and survival of firms – thus, the potential indirect negative impacts on society through lower employment opportunities and lower wealth generation. In particular, empirical research finds that a high level of openness to external knowledge could harm the production of new knowledge and, subsequently, firms' performance (Cruz-González et al. 2015; Veer, Lorenz, and Blind 2016). For instance, Cruz-González et al. (2015), on the basis of a survey data of 248 high-technology manufacturing Spanish firms, find that a high level of openness to external knowledge is positively associated to performance, in more mature sectors, but is harmful to firm performance in more technologically dynamic environments. Also, a recent empirical study by Veer, Lorenz, and Blind (2016), on the cooperative R&D behaviour of 2,792 German firms, shows that all kinds of open innovation collaborations other than with universities and research institutions, are associated to the risk of imitation, regardless of the innovation phase of the firm. Finally, Hopkins et al. (2011) suggest that open innovation needs to be analysed in different sectoral contexts and cannot be considered a panacea for firms wanting to invent. The issues, in these cases, are to find an appropriate balance between internal and external knowledge sources, given the considerable heterogeneity among firms with respect to their collaboration outcomes (Sivadas and Dwyer 2000).

How has the research in these five clusters evolved over time? Table 1 presents the key metrics of the clusters identified through the bibliographic coupling analysis. Based on both total and relative number of citations received in the scientific literature by the articles in the bibliographic sample, we observe that work in Cluster B (*Unsustainable transitions*) is the most frequently cited among the five clusters, and has seen the greatest growth (Figure 4), followed by Cluster A (*Work-related consequences of technology acceptance*), which has a slightly higher number of average citations per article per year (6.84 vs 4.25). A further insight emerges from consideration of the total number of citations received and the growth in the number of articles per year in cluster C (*Innovation & growth downside effects*), which suggests increased attention from innovation scholars and economists working on innovation-related issues. As Table 1 shows, research in the other clusters received far fewer citations and has experienced more limited growth in terms of number of publications per year (Figure 4), indicating less popular research directions.

3.2. Theoretical antecedents: results of the co-citation analysis

To explore the theoretical antecedents of research on the noxious consequences of innovation, we performed co-citation analysis of references and authors. We use the

Table 1. Citation-based statistics of the five clusters (1991 – 2017).

Cluster	Number of articles	Average publication (year)	Total number of citations	Average citations per article	Average number of citations per article per year
A: *Work-related consequences of technology acceptance*	29	2012	992	34.21	6.84
B: *Unsustainable transitions*	36	2010	1,070	29.72	4.25
C: *Innovation & growth downside effects*	19	2013	118	6.21	1.55
D: *The dangers of emerging technologies*	13	2014	90	6.92	2.31
E: *Open innovation's dark side*	10	2012	137	13.70	2.74
Total	107		2,407	21.01	4.72
% dataset	85.60%				

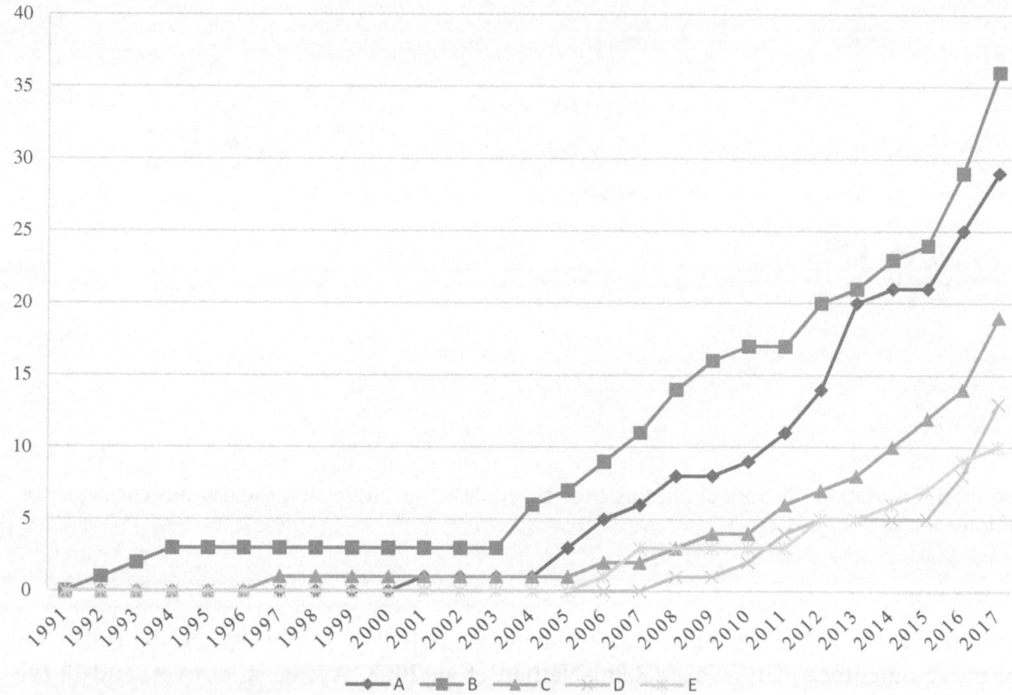

Figure 4. Cumulated number of publication per year per cluster.

Note: Cluster labels are as follows A: *Work-related consequences of technology acceptance*; B: *Unsustainable transitions*; C: *Innovation & growth downside effects*; D: *The dangers of emerging technologies*; E: *Open innovation's dark side*.

Source: Authors' own elaboration.

reference co citation analysis to identify the most cited references and use author co-citation analysis to build a bibliographic network of cited seminal authors (Figure 5).

Our analysis suggests that the most frequently co-cited publications are pioneering articles or chapters in books on innovation such as Nelson and Winter (1982), Dosi (1982), Davis (1989), Cohen and Levinthal (1990), Rogers (1995) and, more recently, Chesbrough (2003) (see Appendix B for the list of the top 10 most influential papers). Beside these seminal contributions, the co-citation analysis allowed us to also identify some specific intellectual antecedents of the five clusters (A to E) by allowing their linking to the most cited authors and references. Cluster A (*Work-related consequences of technology acceptance*) draws mainly on the Technology Acceptance Model (TAM) developed to predict and explain social behaviour. The TAM is grounded in the theory of reasoned actions (Fishbein and Ajzen 1975) and was proposed by Davis (1989) to understand computer usage intention and actual usage behaviour. Others developed the original TAM, to propose TAM2 (Venkatesh and Davis 2000) and the Unified Theory of Acceptance and Use of Technology (UTAUT) (Venkatesh et al. 2003). In broad terms, research on the antecedents of ICT adoption, demonstrates that the intention to use an innovation is the only accurate predictor of its actual adoption and use (Chang and Cheung 2001) while work on the consequences of ICT adoption builds, mainly, on psychometric theory (Nunnally 1978). This strand of work identifies the presence of

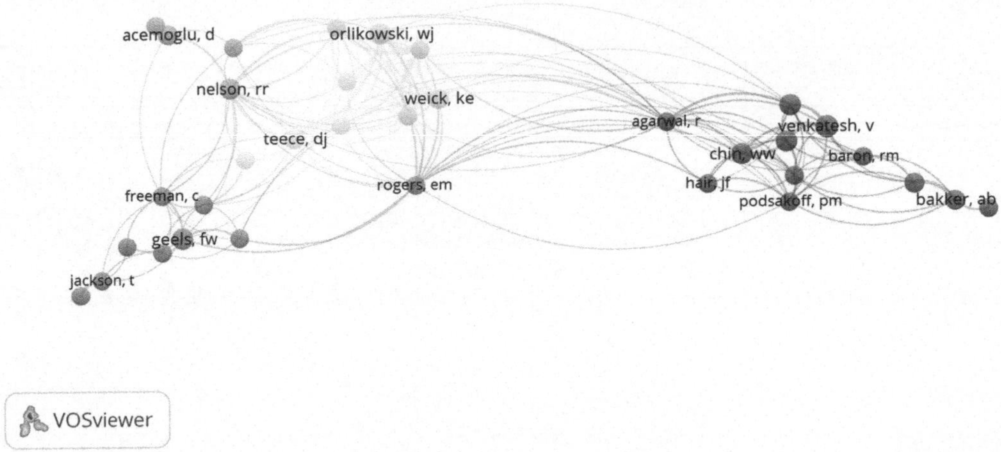

Figure 5. Co-citation network of authors cited by publications addressing the noxious consequences of innovation.
Source: Authors' own elaboration.

chronic stress in the workplace and associated potential inhibiting mechanisms and negative outcomes of ICT use (Ragu-Nathan et al. 2008; Ayyagari, Grover, and Purvis 2011). Moral disengagement theory has been proposed to explain coping mechanisms (D'Arcy et al. 2014), and social cognitive theories have been applied to find ways to counteract the effects of 'technostress' (Tarafdar, Pullins, and Ragu-Nathan 2015). To study ICT-induced interruptions and multitasking behaviour, distraction conflict theory (Baron and Kenny 1986) provides a theoretical perspective to explain the distribution of attention between primary tasks and interruptions. Finally, one of the origins of this scholarly sub-community is Maslach Burnout Inventory (MBI) (Maslach and Jackson 1981) which is used to study work-related burnout.

The second cluster (*Unsustainable transitions*) draws mostly on work on the sustainability of the current production and consumption models in a specific socio-technical system. The baseline theories related to system-level social and technological changes, include transition theory (Geels 2002, 2005; Geels and Schot 2007) and transition management (Rotmans, Kemp, and van Asselt 2001; Kemp and Rotmans 2005). Investigations of socio-technical transition include work on technological innovation and related institutional changes (i.e. policy and regulation, beliefs and values, behavioural expectations, governance structures, learning practices and market structures) (Loorbach 2010).

The third cluster (*Innovation & growth downside effects*) emphasises the influence of technological change and innovation on economic growth and is linked strongly to the evolutionary economics (Dosi 1982; Nelson and Winter 1982) and related management literatures (Henderson and Clark 1990). Some of the research in this cluster is linked, also, to more mainstream economics research on innovation and growth (Grossman and Helpman 1991; Aghion and Howitt 1992). Studies of the harmful effects of innovation exploit these theories to uncover the negative externalities – such as, unemployment, inequality and environmental disasters – resulting from technological change-driven economic growth.

The fourth cluster (*The dangers of emerging technologies*) includes research investigating the relationship between science and technology and society, with particular reference to the broader societal implications of emerging technologies and the role of policymakers for keeping abreast of industry developments. In terms of theory, scholars in this sub-community draw on business ethics (Bunge 1977; Grunwald 1999), the notion of RRI, which tries to align innovation processes and outcomes to societal needs, through the early involvement of a range of actors (e.g. researchers, civil society organisations, industry and policy makers) (e.g. Owen, Macnaghten, and Stilgoe 2012).

Cluster 5 (*Open innovation's dark side*), focuses on how firms' search behaviour influences their ability to innovate and extract the value deriving from innovation and is linked, mostly, to the literature on open innovation and external openness (Chesbrough, Vanhaverbeke, and West 2006; Laursen and Salter 2006). The research in this cluster builds on the core constructs in innovation studies such as Cohen and Levinthal (1990) absorptive capacity and Teece, Pisano, and Shuen (1997) dynamic capabilities. Work on the harmful effects of innovation draws on this literature strand to show how, under certain conditions, open search strategies can harm firm performance and innovativeness.

4. Concluding remarks

Although generally considered a driver of growth, innovation can have noxious consequences for the economy, society and the environment. While most research so far has focused on the upside of innovation, in this review, we show that the academic community is becoming more alert to the downside of innovation. We identified five scholarly research sub-communities examining the *Work-related consequences of technology acceptance* (cluster A); *Unsustainable transitions* (cluster B); *Innovation & growth downside effects* (cluster C); *The dangers of emerging technologies* (cluster D) and the *Open innovation's dark side* (cluster E). It seems that these research agendas have developed largely in parallel and have not converged to become a single field of enquiry, perhaps because each sub-community draws on a different intellectual tradition and develops around different research questions. Nevertheless, overall, these five sub-communities build on and are inspired by a common set of pioneering works, which correspond to the foundation to innovation studies.

We consider that attention to the downside of innovation is timely; it is based on the current unique historical moment and the increased scrutiny and criticism attracted by contemporary capitalism (Giuliani 2018, among many others). It is useful, therefore, to conceptualise the notion of noxious innovation and investigate its origins, nature and consequences. Innovations are not self-regulating entities: their development, use and deployment depend on a range of actors and decision makers, including both regulators and policy-makers, but also company CEOs, top management teams and firm employees who – perhaps unwittingly – make wrong decisions. Also, these actors may be embedded in organisational routines that contribute to the generation or perpetuation of innovative activity's noxious impacts and may do little to mitigate and remedy these effects. In the past, innovation scholars were persuaded by the idea that innovation could have (short term) downside effects, but that these would certainly be offset by innovation-related generation of value for the economy and, consequently, for society as a whole. However, as society and analysts have become more alert to the grand sustainability challenges posed by contemporary capitalism – from climate change to rising populisms and other societal and environmental catastrophes – there is a growing consensus about the need to change that narrative. One way to do that could be to avoid presenting the positive and negative effects of innovation such that the former can offset the latter. Despite efforts made by academic researchers to underline this point, it seems that economic and policy decision makers remain firmly anchored in a very consequentialist narrative which suggests that some harmful impacts are a necessary evil to achieve the 'greatest good'. Changing that narrative will not be easy and might require more concrete scientific evidence, which we hope will be forthcoming.

Against this context, we would suggest two areas requiring further examination. One is the need for innovation studies to advance methods to measure noxious inventions to allow predictions of what will be introduced into the market, based on assessment of the potential toxicity related to granted patents. Might the damage caused by glyphosate and other chemical aberrations, such as DDT, have been avoided had we been better able to predict their potential hazardous impacts on human health and the natural environment? Most such toxicity assessments are conducted in the context of clinical and pharmacological research; however, for public policy reasons, it is important to have better predictive power about the wave of new hazardous chemical substances which might be available on the market and on which companies and other organisations are conducting R&D. We think it would be useful to develop new measures to capture the potentially harmful impacts of patented inventions, which, so far, have been analysed based on consideration of their valuable characteristics – in terms of impact and quality. For instance, Biggi, Giuliani, and Martinelli (2019) propose the use of computational chemistry techniques to assess the potential noxious effects of chemical and pharmaceutical technologies on human health, the environment and ecosystems, by analysing the groups of compounds disclosed in patent data. This type of research might open the way to a new field of scholarly investigation on the temporal geography of potentially hazardous inventions, to assess whether, when, where and why they manage to come to market, who are the big market players and which governments are facilitating their diffusion.

Second, research on noxious innovations so far, examines either meso to macro phenomena (as in the case of cluster B, C or D) or some micro dimensions regarding individuals (as in cluster A), but mostly overlooks the role that companies and their strategic decisions play, in shaping the harmful impacts of innovations on society – the work in cluster E touches on this only tangentially by discussing open innovation. However, innovating companies – especially big global players – have great responsibility. There is a growing consensus that shareholder capitalism does not maximise welfare and this, as Joseph Stiglitz recently and rightly argued, is manifest when corporations 'poison the air we breathe or the water we drink ... ; push unhealthy products like sugary drinks that contribute to childhood obesity, or painkillers that unleash an opioid crisis'.[7] As innovation scholars, we need to look more closely at the contexts of these corporations' decisions to continue to provide services and products that have important negative consequences on the environment and society.

[7] https://www.project-syndicate.org/.

Finally, we would highlight some limitations of this study. Although use of bibliometric techniques should have reduced some subjective bias typical of traditional reviews (Kovacs, Van Looy, and Cassiman 2015), our work should be interpreted with some caution. First, we acknowledge that we have opted for a top down selection of keywords, which may have led to the exclusion of some work on phenomena that conceptually are similar to the noxious consequences of innovation. We hope that future research will be able to identify new terminologies to continue mapping this literature and its evolution over time. Second, the use of bibliographic databases, such as WoS and Scopus, might have introduced a bias towards scholarly work from countries that publish in English language journals (Mongeon and Paul-Hus 2016) and, by being more selective than other datasets (see, e.g. Google Scholar or the DOAJ – Directory of Open Access Journals), may have underestimated the extent of the research in each individual sub-community or failed to appreciate the emergence of new topics which are yet be included in WoS/Scopus indexed journals. Finally, the citations analysis assumes that citations are of substantive importance to represent scholarly work and scholarly communities (King 1987; Campbell et al. 2010), but given our reliance on WoS and Scopus, we are able to measure the citations by papers included in such datasets, but not those coming from other sources. Thus, this review and our analyses should be seen as complementing alternative fine-grained content analyses based on extensive reading (Schraven, Hartmann, and Dewulf 2015; White and McCain 1998).

Disclosure statement

No potential conflict of interest was reported by the authors.

References

Agarwal, R., and J. Prasad. 1998. "A Conceptual and Operational Definition of Personal Innovativeness in the Domain of Information Technology." *Information Systems Research* 9 (2): 204–215. doi:10.1287/isre.9.2.204.

Aghion, P., and P. Howitt. 1992. "A Model of Growth through Creative Destruction." *Econometrica* 60 (2): 322–352. doi:10.2307/2951599.

Aghion, P., and P. Howitt. 1998. *Endogenous Growth Theory*. Cambridge, MA: MIT Press.

Ayyagari, R., V. Grover, and R. Purvis. 2011. "Technostress: Technological Antecedents and Implications." *MIS Quarterly* 35 (4): 831–858. doi:10.2307/41409963.

Baron, R. M., and D. A. Kenny. 1986. "The Moderator–Mediator Variable Distinction in Social Psychological Research: Conceptual, Strategic, and Statistical Considerations." *Journal of Personality and Social Psychology* 51 (6): 1173–1182. doi:10.1037/0022-3514.51.6.1173.

Biggi, G., E. Giuliani, and A. Martinelli. 2019. "Patent Toxicity: An Exploratory Comparative Analysis on the Noxious Impact of Chemical Inventions." Paper presented at the annual meeting of the Academy of Management, Boston, August 9–11.

Börner, K., C. Chen, and K. W. Boyack. 2003. "Visualizing Knowledge Domains." *Annual Review of Information Science and Technology* 37 (1): 179–255. doi:10.1002/aris.1440370106.

Boyack, K. W., and R. Klavans. 2010. "Co-Citation Analysis, Bibliographic Coupling, and Direct Citation: Which Citation Approach Represents the Research Front Most Accurately?" *Journal of the American Society for Information Science and Technology* 61 (12): 2389–2404. doi:10.1002/asi.21419.

Breau, S., D. F. Kogler, and K. C. Bolton. 2014. "On the Relationship between Innovation and Wage Inequality: New Evidence from Canadian Cities." *Economic Geography* 90 (4): 351–373. doi:10.1111/ecge.12056.

Bunge, M. 1977. "Towards a Technoethics." *Monist* 60 (1): 96–107. doi:10.5840/monist197760134.

Campbell, D., M. Picard-Aitken, G. Côté, J. Caruso, R. Valentim, S. Edmonds, and M. C. Laframboise. 2010. "Bibliometrics as a Performance Measurement Tool for Research Evaluation: The Case of Research Funded by the National Cancer Institute of Canada." *American Journal of Evaluation* 31 (1): 66–83. doi:10.1177/1098214009354774.

Chang, M. K., and W. Cheung. 2001. "Determinants of the Intention to Use Internet/WWW at Work: A Confirmatory Study." *Information & Management* 39 (1): 1–14. doi:10.1016/S0378-7206(01)00075-1.

Chesbrough, H. W. 2003. *Open Innovation: The New Imperative for Creating and Profiting from Technology*. Boston, MA: Harvard Business Press.

Chesbrough, H. W., W. Vanhaverbeke, and J. West. 2006. *Open Innovation: Researching a New Paradigm*. Oxford: Oxford University Press.

Cobo, M. J., A. G. López-Herrera, E. Herrera-Viedma, and F. Herrera. 2011. "Science Mapping Software Tools: Review, Analysis, and Cooperative Study among Tools." *Journal of the American Society for Information Science and Technology* 62 (7): 1382–1402. doi:10.1002/asi.21525.

Cohen, W. M., and D. A. Levinthal. 1990. "Absorptive Capacity: A New Perspective on Learning and Innovation." *Administrative Science Quarterly* 35 (1): 128–152. doi:10.2307/2393553.

Crane, D. 1972. *Diffusion of Knowledge in Scientific Communities*. Chicago: University of Chicago Press.

Cruz-González, J., P. López-Sáez, J. E. Navas-López, and M. Delgado-Verde. 2015. "Open Search Strategies and Firm Performance: The Different Moderating Role of Technological Environmental Dynamism." *Technovation* 35: 32–45. doi:10.1016/j.technovation.2014.09.001.

D'Arcy, J., A. Gupta, M. Tarafdar, and O. Turel. 2014. "Reflecting on the "Dark Side" of Information Technology Use." *Communications of the Association for Information Systems* 35: 109–118. doi:10.17705/1CAIS.03505.

Davis, F. D. 1989. "Perceived Usefulness, Perceived Ease of Use, and User Acceptance of Information Technology." *MIS Quarterly* 13 (3): 319–340. doi:10.2307/249008.

Di Maria, C., and S. A. Smulders. 2004. "Trade Pessimists Vs Technology Optimists: Induced Technical Change and Pollution Havens." *Advances in Economic Analysis & Policy* 3 (2): 1–27. doi:10.2202/1538-0637.1344.

Dilaver, O. 2014. "Involuntary Technology Adoptions: How Consumer Interdependencies Lead to Societal Change." *Structural Change and Economic Dynamics* 31: 138–148. doi:10.1016/j.strueco.2014.09.003.

Dosi, G. 1982. "Technological Paradigms and Technological Trajectories: A Suggested Interpretation of the Determinants and Directions of Technical Change." *Research Policy* 11 (3): 147–162. doi:10.1016/0048-7333(82)90016-6.

Dosi, G. 2011. "Economic Coordination and Dynamics: Some Elements of an Alternative "Evolutionary" Paradigm." Accessed 31 December 2019. http://www2.econ.iastate.edu/tesfatsi/EconomicCoordinationAndDynamics.GDosi2012.INET.pdf

Dosi, G. 2013. "Innovation, Evolution, and Economics: Where We are and Where We Should Go." In *Innovation Studies: Evolution and Future Challenges*, edited by J. Fagerberg, B. E. Martin, and E. S. Andersen, 111–133. Oxford: Oxford University Press.

Dosi, G., C. Freeman, R. Nelson, G. Silverberg, and L. Soete. 1988. *Technical Change and Economic Theory*. London: Pinter Publishers.

Eisenhardt, K. M. 1989. "Agency Theory: An Assessment and Review." *Academy of Management Review* 14 (1): 57–74. doi:10.5465/amr.1989.4279003.

Fishbein, M., and I. Ajzen. 1975. *Belief, Attitude, Intention and Behavior: An Introduction to Theory and Research*. Reading, MA: Addison-Wesley.

Freeman, C. 1988. "Japan: A New National System of Innovation." In *Technical Change and Economic Theory*, edited by G. Dosi, C. Freeman, R. R. Nelson, G. Silverberg, and L. Soete, 330–348. London: Pinter Publishers.

Frondel, M. 2004. "Energy Conservation, the Rebound Effect, and Future Energy and Transport Technologies: An Introduction to Energy Conservation and the Rebound Effect." *International Journal of Energy Technology and Policy* 2 (3): 203–208.

Garfield, E., M. V. Malin, and H. Small. 1983. "Citation Data as Science Indicators." In *Toward a Metric of Science: The Advent of Science Indicators*, edited by Y. Elkana, J. Ledelberg, R. K. Merton, A. Thackray, and A. Zuckerman, 580–608. New York: John Wiley and Sons.

Geels, F. W. 2002. "Technological Transitions as Evolutionary Reconfiguration Processes: A Multi-Level Perspective and A Case-Study." *Research Policy* 31 (8–9): 1257–1274. doi:10.1016/S0048-7333(02)00062-8.

Geels, F. W. 2004. "From Sectoral Systems of Innovation to Socio-Technical Systems: Insights about Dynamics and Change from Sociology and Institutional Theory." *Research Policy* 33 (6–7): 897–920. doi:10.1016/j.respol.2004.01.015.

Geels, F. W. 2005. *Technological Transitions and System Innovations: A Co-Evolutionary and Socio-Technical Analysis*. Cheltenham: Edward Elgar Publishing.

Geels, F. W., and J. Schot. 2007. "Typology of Sociotechnical Transition Pathways." *Research Policy* 36 (3): 399–417. doi:10.1016/j.respol.2007.01.003.

Giuliani, E. 2018. "Regulating Global Capitalism amid Rampant Corporate Wrongdoing-Reply to "Three Frames for Innovation Policy." *Research Policy* 47 (9): 1577–1582. doi:10.1016/j.respol.2018.08.013.

Golden, T. D. 2006. "Avoiding Depletion in Virtual Work: Telework and the Intervening Impact of Work Exhaustion on Commitment and Turnover Intentions." *Journal of Vocational Behavior* 69 (1): 176–187. doi:10.1016/j.jvb.2006.02.003.

Golden, T. D., J. F. Veiga, and R. N. Dino. 2008. "The Impact of Professional Isolation on Teleworker Job Performance and Turnover Intentions: Does Time Spent Teleworking, Interacting Face-to-Face, or Having Access to Communication-Enhancing Technology Matter?" *Journal of Applied Psychology* 93 (6): 1412–1421. doi:10.1037/a0012722.

Griggs, D., M. Stafford-Smith, O. Gaffney, J. Rockström, M. C. Öhman, P. Shyamsundar, W. Steffen, et al. 2013. "Policy: Sustainable Development Goals for People and Planet." *Nature* 495 (7441): 305–307. doi:10.1038/495305a.

Grossman, G. M., and E. Helpman. 1991. "Trade, Knowledge Spillovers, and Growth." *European Economic Review* 35 (2–3): 517–526. doi:10.1016/0014-2921(91)90153-A.

Groves, C., L. Frater, R. Lee, and E. Stokes. 2011. "Is There Room at the Bottom for CSR? Corporate Social Responsibility and Nanotechnology in the UK." *Journal of Business Ethics* 101 (4): 525–552. doi:10.1007/s10551-010-0731-7.

Grunwald, A. 1999. "Technology Assessment or Ethics of Technology? Reflections on Technology Development between Social Sciences and Philosophy." *Ethical Perspectives* 6 (2): 170–182. doi:10.2143/ep.6.2.505355.

Hellström, T. 2003. "Systemic Innovation and Risk: Technology Assessment and the Challenge of Responsible Innovation." *Technology in Society* 25 (3): 369–384. doi:10.1016/S0160-791X(03)00041-1.

Henderson, B. J., and N. D. Pearson. 2011. "The Dark Side of Financial Innovation: A Case Study of the Pricing of A Retail Financial Product." *Journal of Financial Economics* 100 (2): 227–247. doi:10.1016/j.jfineco.2010.12.006.

Henderson, R. M., and K. B. Clark. 1990. "Architectural Innovation: The Reconfiguration of Existing Product Technologies and the Failure of Established Firms." *Administrative Science Quarterly* 35 (1): 9–30. doi:10.2307/2393549.

Herring, H. 2004. "The Rebound Effect and Energy Conservation." In *The Encyclopaedia of Energy*, edited by C. Cleveland, 411–423. Amsterdam: Academic Press.

Herring, H., and R. Roy. 2007. "Technological Innovation, Energy Efficient Design and the Rebound Effect." *Technovation* 27 (4): 194–203. doi:10.1016/j.technovation.2006.11.004.

Hopkins, M. M., J. Tidd, P. Nightingale, and R. Miller. 2011. “Generative and Degenerative Interactions: Positive and Negative Dynamics of Open, User-Centric Innovation in Technology and Engineering Consultancies.” *R&D Management* 41 (1): 44–60. doi:10.1111/j.1467-9310.2010.00631.x.

Jenkins, R. 2008. “Trade, Technology and Employment in South Africa.” *The Journal of Development Studies* 44 (1): 60–79. doi:10.1080/00220380701722308.

Kemp, R., and J. Rotmans. 2005. “The Management of the Co-Evolution of Technical, Environmental and Social Systems.” In *Towards Environmental Innovation Systems*, edited by M. Weber, and J. Hemmelskamp, 33–55. Berlin: Springer.

Kessler, M. M. 1963. “Bibliographic Coupling Between Scientific Papers.” *American Documentation* 14 (1): 10–25. doi:10.1002/asi.5090140103.

King, J. 1987. “A Review of Bibliometric and Other Science Indicators and Their Role in Research Evaluation.” *Journal of Information Science* 13 (5): 261–276. doi:10.1177/016555158701300501.

Klingelhöfer, H. E. 2017. “Unintended Possible Consequences of Fuel Input Taxes for Individual Investments in Greenhouse Gas Mitigation Technologies and the Resulting Emissions.” *South African Journal of Economic and Management Sciences* 20 (1): 1–11. doi:10.4102/sajems.v20i1.1472.

Korinek, A., and J. E. Stiglitz. 2017. “Artificial Intelligence and Its Implications for Income Distribution and Unemploymnet.” *NBER Working Paper no. 24174*. Cambridge, MA: National Bureau for Economic Research.

Kovacs, A., B. Van Looy, and B. Cassiman. 2015. “Exploring the Scope of Open Innovation: A Bibliometric Review of A Decade of Research.” *Scientometrics* 104 (3): 951–983. doi:10.1007/s11192-015-1628-0.

Laursen, K., and A. Salter. 2006. “Open for Innovation: The Role of Openness in Explaining Innovation Performance among UK Manufacturing Firms.” *Strategic Management Journal* 27 (2): 131–150. doi:10.1002/smj.507.

Lee, N., and A. Rodríguez-Pose. 2013. “Innovation and Spatial Inequality in Europe and USA.” *Journal of Economic Geography* 13 (1): 1–22. doi:10.1093/jeg/lbs022.

Lohmann, R., K. Breivik, J. Dachs, and D. Muir. 2007. “Global Fate of POPs: Current and Future Research Directions.” *Environmental Pollution* 150 (1): 150–165. doi:10.1016/j.envpol.2007.06.051.

Loorbach, D. 2010. “Transition Management for Sustainable Development: A Prescriptive, Complexity-Based Governance Framework.” *Governance* 23 (1): 161–183. doi:10.1111/j.1468-0491.2009.01471.x.

Lu, L. Y., B. J. Lin, J. S. Liu, and C. Y. Yu. 2012. “Ethics in Nanotechnology: What’s Being Done? What’s Missing?” *Journal of Business Ethics* 109 (4): 583–598. doi:10.1007/s10551-012-1432-1.

Lundvall, B. Å. 1992. *National Systems of Innovation: Towards a Theory of Innovation and Interactive Learning*. London: Pinter Publishers.

Luzzati, T., A. Parenti, and T. Rughi. 2018. “Economic Growth and Cancer Incidence.” *Ecological Economics* 146: 381–396. doi:10.1016/j.ecolecon.2017.11.031.

Managi, S., and S. Kumar. 2009. “Trade-Induced Technological Change: Analyzing Economic and Environmental Outcomes.” *Economic Modelling* 26 (3): 721–732. doi:10.1016/j.econmod.2009.02.002.

Markard, J., R. Raven, and B. Truffer. 2012. “Sustainability Transitions: An Emerging Field of Research and Its Prospects.” *Research Policy* 41 (6): 955–967. doi:10.1016/j.respol.2012.02.013.

Martyn, J. 1964. “Bibliographic Coupling.” *Journal of Documentation* 20 (4): 236. doi:10.1108/eb026352.

Maslach, C., and S. E. Jackson. 1981. “The Measurement of Experienced Burnout.” *Journal of Occupational Behavior* 2 (2): 99–113. doi:10.1002/job.4030020205.

Maslach, C., W. B. Schaufeli, and M. P. Leiter. 2001. “Job Burnout.” *Annual Review of Psychology* 52 (1): 397–422. doi:10.1146/annurev.psych.52.1.397.

Meadowcroft, J. 2011. “Engaging with the Politics of Sustainability Transitions.” *Environmental Innovation and Societal Transitions* 1 (1): 70–75. doi:10.1016/j.eist.2011.02.003.

Merton, R. K. 1936. "The Unanticipated Consequences of Purposive Social Action." *American Sociological Review* 1 (6): 894–904. doi:10.2307/2084615.

Meyer, M., K. Grant, P. Morlacchi, and D. Weckowska. 2014. "Triple Helix Indicators as an Emergent Area of Enquiry: A Bibliometric Perspective." *Scientometrics* 99 (1): 151–174. doi:10.1007/s11192-013-1103-8.

Mongeon, P., and A. Paul-Hus. 2016. "The Journal Coverage of Web of Science and Scopus: A Comparative Analysis." *Scientometrics* 106 (1): 213–228. doi:10.1007/s11192-015-1765-5.

Nelson, R. R., and S. G. Winter. 1982. *An Evolutionary Theory of Economic Change*. Cambridge, MA: Beknap Harvard.

Newman, M. E., and M. Girvan. 2004. "Finding and Evaluating Community Structure in Networks." *Physical Review E* 69 (2): 026113. doi:10.1103/PhysRevE.69.026113.

Noyons, E. C., H. F. Moed, and M. Luwel. 1999. "Combining Mapping and Citation Analysis for Evaluative Bibliometric Purposes: A Bibliometric Study." *Journal of the American Society for Information Science* 50 (2): 115–131. doi:10.1002/(SICI)1097-4571(1999)50:2<115::AID-ASI3>3.0.CO;2-J.

Nunnally, J. C. 1978. *Psychometric Theory*. New York: McGraw-Hill.

Owen, R., P. Macnaghten, and J. Stilgoe. 2012. "Responsible Research and Innovation: From Science in Society to Science for Society, with Society." *Science and Public Policy* 39 (6): 751–760. doi:10.1093/scipol/scs093.

Pérignon, C., and B. Vallée. 2017. "The Political Economy of Financial Innovation: Evidence from Local Governments." *The Review of Financial Studies* 30 (6): 1903–1934. doi:10.1093/rfs/hhx029.

Podsakoff, P. M., S. B. MacKenzie, J. Y. Lee, and N. P. Podsakoff. 2003. "Common Method Biases in Behavioral Research: A Critical Review of the Literature and Recommended Remedies." *Journal of Applied Psychology* 88 (5): 879–903. doi:10.1037/0021-9010.88.5.879.

Raasch, C., V. Lee, S. Spaeth, and C. Herstatt. 2013. "The Rise and Fall of Interdisciplinary Research: The Case of Open Source Innovation." *Research Policy* 42 (5): 1138–1151. doi:10.1016/j.respol.2013.01.010.

Ragu-Nathan, T. S., M. Tarafdar, B. S. Ragu-Nathan, and Q. Tu. 2008. "The Consequences of Technostress for End Users in Organizations: Conceptual Development and Empirical Validation." *Information Systems Research* 19 (4): 417–433. doi:10.1287/isre.1070.0165.

Roca, J. B., P. Vaishnav, M. G. Morgan, J. Mendonça, and E. Fuchs. 2017. "When Risks Cannot Be Seen: Regulating Uncertainty in Emerging Technologies." *Research Policy* 46 (7): 1215–1233. doi:10.1016/j.respol.2017.05.010.

Rockström, J., W. Steffen, K. Noone, Å. Persson, F. S. Chapin, I. I. I., . E. Lambin, T. M. Lenton, et al. 2009. "Planetary Boundaries: Exploring the Safe Operating Space for Humanity." *Nature* 461 (7263): 472–475. doi:10.1038/461472a.

Rogers, E. M. 1995. *Diffusion of Innovations: Modifications of a Model for Telecommunications*. Berlin, Heidelberg: Springer.

Romer, P. M. 1986. "Increasing Returns and Long-run Growth." *Journal of Political Economy* 94 (5): 1002–1103. doi:10.1086/261420.

Røpke, I., T. H. Christensen, and J. O. Jensen. 2010. "Information and Communication Technologies – A New Round of Household Electrification." *Energy Policy* 38 (4): 1764–1773. doi:10.1016/j.enpol.2009.11.052.

Rotmans, J., R. Kemp, and M. van Asselt. 2001. "More Evolution than Revolution: Transition Management in Public Policy." *Foresight* 3 (1): 15–31. doi:10.1108/14636680110803003.

Roy, U. 1997. "Economic Growth with Negative Externalities in Innovation." *Journal of Macroeconomics* 19 (1): 155–173. doi:10.1016/s0164-0704(97)00009-8.

Schildt, H. A., S. A. Zahra, and A. Sillanpää. 2006. "Scholarly Communities in Entrepreneurship Research: A Co–citation Analysis." *Entrepreneurship Theory and Practice* 30 (3): 399–415. doi:10.1111/j.1540-6520.2006.00126.x.

Schot, J., and W. E. Steinmueller. 2018. "Three Frames for Innovation Policy: R&D, Systems of Innovation and Transformative Change." *Research Policy* 47 (9): 1554–1567. doi:10.1016/j.respol.2018.08.011.

Schraven, D. F., A. Hartmann, and G. P. Dewulf. 2015. "Research Orientations Towards the 'Management'of Infrastructure Assets: An Intellectual Structure Approach." *Structure and Infrastructure Engineering* 11 (2): 73–96. doi:10.1080/15732479.2013.848909.

Schumpeter, J. A. 1942. *Capitalism, Socialism and Democracy*. New York: Harper.

Sivadas, E., and F. R. Dwyer. 2000. "An Examination of Organizational Factors Influencing New Product Success in Internal and Alliance-Based Processes." *Journal of Marketing* 64 (1): 31–49. doi:10.1509/jmkg.64.1.31.17985.

Soete, L. 2013. "Is Innovation Always Good?" In *The Future of InnovationStudies: Evolution and Future Challenges*, edited by J. Fagerberg, B. R. Martin, and E. S. Andersen, 134–144. Oxford: Oxford University Press.

Steffen, W., P. J. Crutzen, and J. R. McNeill. 2007. "The Anthropocene: Are Humans Now Overwhelming the Great Forces of Nature." *AMBIO: A Journal of the Human Environment* 36 (8): 614–621. doi:10.1579/0044-7447(2007)36[614:TAAHNO]2.0.CO;2.

Stilgoe, J., R. Owen, and P. Macnaghten. 2013. "Developing a Framework for Responsible Innovation." *Research Policy* 42 (9): 1568–1580. doi:10.1016/j.respol.2013.05.008.

Tarafdar, M., E. B. Pullins, and T. S. Ragu-Nathan. 2015. "Technostress: Negative Effect on Performance and Possible Mitigations." *Information Systems Journal* 25 (2): 103–132. doi:10.1111/isj.12042.

Teece, D. J., G. Pisano, and A. Shuen. 1997. "Dynamic Capabilities and Strategic Management." *Strategic Management Journal* 18 (7): 509–533. doi:10.1002/(sici)1097-0266(199708)18:7<509::aid-smj882>3.0.co;2-z.

Thyer, B. A. 2008. *Preparing Research Articles*. Oxford: Oxford University Press.

Tranfield, D., D. Denyer, and P. Smart. 2003. "Toward a Methodology for Developing Evidence-Informed Management Knowledge by Means of Systematic Review." *British Journal of Management* 14: 207–222. doi:10.1111/1467-8551.00375.

Turel, O., A. Serenko, and P. Giles. 2011. "Integrating Technology Addiction and Use: An Empirical Investigation of Online Auction Users." *MIS Quarterly* 35 (4): 1043–1061. doi:10.2307/41409972.

United Nations. 2014. "The Ten Principles." Accessed 3 January 2019 https://www.globalcompact.ca/about/ungc-10-principles/

United Nations. 2015. "Transforming Our World: The 2030 Agenda for Sustainable Development." October 21. Accessed 3 January 2019 https://www.refworld.org/docid/57b6e3e44.html

van der Have, R. P., and L. Rubalcaba. 2016. "Social Innovation Research: An Emerging Area of Innovation Studies?" *Research Policy* 45 (9): 1923–1935. doi:10.1016/j.respol.2016.06.010.

van Eck, N. J., and L. Waltman. 2009. "How to Normalize Cooccurrence Data? an Analysis of Some Well-Known Similarity Measures." *Journal of the American Society for Information Science and Technology* 60 (8): 1635–1651. doi:10.1002/asi.21075.

van Eck, N. J., and L. Waltman. 2010. "Software Survey: VOSviewer, a Computer Program for Bibliometric Mapping." *Scientometrics* 84 (2): 523–538. doi:10.1007/s11192-009-0146-3.

van Oorschot, J. A., E. Hofman, and J. I. Halman. 2018. "A Bibliometric Review of the Innovation Adoption Literature." *Technological Forecasting and Social Change* 134: 1–21. doi:10.1016/j.techfore.2018.04.032.

van Zoonen, W., J. W. Verhoeven, and R. Vliegenthart. 2016. "Social Media's Dark Side: Inducing Boundary Conflicts." *Journal of Managerial Psychology* 31 (8): 1297–1311. doi:10.1108/JMP-10-2015-0388.

Veer, T., A. Lorenz, and K. Blind. 2016. "How Open Is Too Open? the Mitigating Role of Appropriation Mechanisms in R&D Cooperation Settings." *R&D Management* 46 (S3): 1113–1128. doi:10.1111/radm.12232.

Venkatesh, V., and F. D. Davis. 2000. "A Theoretical Extension of the Technology Acceptance Model: Four Longitudinal Field Studies." *Management Science* 46 (2): 186–204. doi:10.1287/mnsc.46.2.186.11926.

Venkatesh, V., M. G. Morris, G. B. Davis, and F. D. Davis. 2003. “User Acceptance of Information Technology: Toward a Unified View.” *MIS Quarterly* 27 (3): 425–478. doi:10.2307/30036540.

Vogel, R., and W. H. Güttel. 2013. “The Dynamic Capability View in Strategic Management: A Bibliometric Review.” *International Journal of Management Reviews* 15 (4): 426–446. doi:10.1111/ijmr.12000.

Waltman, L., and N. J. Van Eck. 2013. “A Smart Local Moving Algorithm for Large-Scale Modularity-Based Community Detection.” *The European Physical Journal B* 86 (11): 1–14. doi:10.1140/epjb/e2013-40829-0.

Wang, H., P. Zhou, and D. Q. Zhou. 2012. “An Empirical Study of Direct Rebound Effect for Passenger Transport in Urban China.” *Energy Economics* 34 (2): 452–460. doi:10.1016/j.eneco.2011.09.010.

West, J., and M. Bogers. 2014. “Leveraging External Sources of Innovation: A Review of Research on Open Innovation.” *Journal of Product Innovation Management* 31 (4): 814–831. doi:10.1111/jpim.12125.

White, H. D., and K. W. McCain. 1998. “Visualizing a Discipline: An Author Co-citation Analysis of Information Science, 1972–1995.” *Journal of the American Society for Information Science* 49 (4): 327–355. doi:10.1002/(SICI)1097-4571(19980401)49:4<327::AID-ASI4>3.0.CO;2-4.

Wigboldus, S., L. Klerkx, C. Leeuwis, M. Schut, S. Muilerman, and H. Jochemsen. 2016. “Systemic Perspectives on Scaling Agricultural Innovations. A Review.” *Agronomy for Sustainable Development* 36 (3): 1–20. doi:10.1007/s13593-016-0380-z.

Wilburn, K. M., and H. R. Wilburn. 2016. “Asking “What Else?” To Identify Unintended Negative Consequences.” *Business Horizons* 59 (2): 213–221. doi:10.1016/j.bushor.2015.11.006.

Youtie, J., L. Kay, and J. Melkers. 2013. “Bibliographic Coupling and Network Analysis to Assess Knowledge Coalescence in a Research Center Environment.” *Research Evaluation* 22 (3): 145–156. doi:10.1093/reseval/rvt002.

Appendix A. Search Strings

(‘dark innovation’ [All Fields]) (‘dark side of innovation’ [All Fields]) (‘destructive technolog*’ [All Fields]) (‘destructive innovat*’ [All Fields]) (‘irresponsible innovation’ [All Fields]) (‘irresponsible’ AND ‘innovat*’ [All Fields])(‘innovat*’ AND 'environmental disaster’ [All Fields]) (‘tech*’ AND ‘environmental disaster’ [All Fields])(‘harm’ AND ‘innovation’ [All Fields])(‘innovat*’ AND ‘negative indirect effect’ [All Fields]) (‘technolog*’ AND ‘negative indirect effect’ [All Fields]) (‘dark side’ AND ‘social media’ [All Fields]) (‘dark side’ AND ‘emerging technolog*’ [All Fields]) (‘dark side’ AND ‘new technolog*’ [All Fields]) (‘ICT’ AND ‘negative impact’ [All Fields]) (‘ICT’ AND ‘negative consequences’ [All Fields])(“contested innovat* [All Fields]) (‘contested technolog*’ [All Fields]) (‘dark matter’ AND ‘innovat*’ [All Fields]) (‘dark matter’ AND ‘technolog*’ [All Fields]) (‘bad innovat*’ [All Fields]) (‘negative externalities’ AND ‘innovat*’ [All Fields]) (‘undesirable effect’ AND ‘innovat*’ [All Fields]) (‘undesirable effect’ AND “technolog* [All Fields]) (‘harmful technolog*’ [All Fields]) (‘harmful technologies’ [All Fields]) (‘harm’ AND ‘technological innovation’ [All Fields]) (‘harm’ AND ‘technolog*’ AND ‘innovat*’ [All Fields]) (‘harm’ AND ‘technolog*’ AND ‘environment’ [All Fields]) (‘harm’ AND ‘technolog*’ AND ‘social’ [All Fields]) (‘hazardous’ AND ‘innovat*’ [All Fields]) (‘hazardous’ AND ‘technolog*’ [All Fields]) (‘harm’ AND technolog* [in Title]) (‘innovat*’ [in Title] AND ‘dark side’ [All Fields]) (‘technolog*’ [in Title] AND ‘dark side’ [All Fields]) (‘innovat*’ AND ‘harm’ AND ‘social’ AND ‘environment*’ [All Fields]) (‘innovat*’ [in Title] AND ‘danger*’ [All Fields] AND ‘social’ [All Fields] AND ‘environment*’ [All Fields]) (‘innovat*’ [in Title] AND ‘negative’ [All Fields] AND ‘unintended’ [All Fields] AND ‘consequences’ [All Fields]) (‘innovat*’ [in Title] AND ‘negative’ [All Fields] AND ‘unwanted’ [All Fields] AND ‘consequences’ [All Fields]) (‘innovat*’ [in Title] AND ‘negative’ [All Fields] AND ‘undesired’ [All Fields] AND ‘consequences’ [All Fields]) (‘technolog*’ [in Title] AND ‘negative’ [All Fields] AND ‘unwanted’ [All Fields] AND ‘consequences’ [All Fields]) (‘technolog*’ [in Title] AND ‘negative’ [All Fields] AND ‘unintended’ [All Fields] AND 'consequences' [All Fields]) ('technolog*' [in Title] AND 'negative' [All Fields] AND 'undesired' [All Fields] AND 'consequences' [All Fields]) ('innovat*' [in Title] AND 'environment*' [All Fields] AND 'disaster' [All Fields]) ('innovat*' [in Title] AND 'social*' [All Fields] AND 'disaster' [All Fields]) ('technolog*' [in Title] AND 'environment*' [All Fields] AND 'disaster' [All Fields]) ('technolog*' [in Title] AND 'social' [All Fields] AND 'disaster' [All Fields]) ('innovat*' [in Title] AND 'negative' [All Fields] AND 'side effect' [All Fields]) ('technolog*' [in Title] AND 'negative' [All Fields] AND 'side effect' [All Fields]) ('innovation' [in Title] AND 'risk' [All Fields]) ('innovat*' [in Title] AND 'negative outcomes' [All Fields]) ('technolog*' [in Title] AND 'negative outcomes' [All Fields]) ('innovat*' [in Title] AND 'undesirable' AND 'outcomes' [All Fields]) ('technolog*' [in Title] AND 'undesirable' AND 'outcomes' [All Fields]) ('innovat*' [in Title] AND 'undesirable' AND 'effect' [All Fields]) ('technolog*' [in Title] AND 'undesirable' AND 'effect' [All Fields]) ('innovat*' [in Title] AND 'indirect effect' [All Fields])('technolog*' [in Title] AND 'indirect effect' [All Fields]) ('innovat*' [in Title] AND 'unexpected' AND 'consequences' [All Fields]) ('technolog*' [in Title] AND 'unexpected' AND 'consequences' [All Fields]) ('innovat*' [in Title] AND 'unforeseen consequences' [All Fields]) ('technolog*' [in Title] AND 'unforeseen consequences' [All Fields]) ('new product' [in Title] AND 'pollution' [All Fields]) ('new product' [in Title] AND 'environmental disaster' [All Fields]) ('new product' [in Title] AND 'negative consequences' [All Fields]) ('new product' [in Title] AND 'unwanted consequences' [All Fields]) ('new product' [in Title] AND 'undesired consequences' [All Fields]) ('new product' [in Title] AND 'unintended consequences' [All Fields])

Appendix B. Most co-cited references – influential papers

Rank	Top co-cited publications (references)
1	Cohen and Levinthal (1990). Absorptive capacity: A new perspective on learning and innovation. *Administrative Science Quarterly*, 128–152.
2	Nelson and Winter (1982). *An evolutionary theory of technical change*. Cambridge, Ma: Beknap Harvard.
3	Davis (1989) Perceived usefulness, perceived ease of use, and user acceptance of information technology. *MIS Quarterly*, 319–340.
4	Podsakoff et al. (2003). Common method biases in behavioural research: A critical review of the literature and recommended remedies. *Journal of Applied Psychology*, 88(5), 879.
5	Rogers (1995) *Diffusion of Innovations: modifications of a model for telecommunications*. Berlin, Heidelberg: Springer.
6	Agarwal and Prasad (1998). A conceptual and operational definition of personal innovativeness in the domain of information technology. *Information Systems Research*, 9(2), 204–215.
7	Chesbrough (2003). *Open innovation: The new imperative for creating and profiting from technology*. Cambridge, MA: Harvard Business Press.
8	Dosi (1982). Technological paradigms and technological trajectories: a suggested interpretation of the determinants and directions of technical change. *Research Policy*, 11(3), 147–162.
9	Eisenhardt (1989). Agency theory: An assessment and review. *Academy of Management Review*, 14(1), 57–74.
10	Henderson and Clark (1990). Architectural innovation: The reconfiguration of existing product technologies and the failure of established firms. *Administrative Science Quarterly*, 9–30.

The dark side of the sun: solar e-waste and environmental upgrading in the off-grid solar PV value chain

Ulrich Elmer Hansen, Ivan Nygaard and Mirko Dal Maso

ABSTRACT

The expected increase in the disposal of off-grid solar e-waste (SEW) in the Global South is potentially the dark side of a promising innovation. We suggest placing research aimed at identifying appropriate schemes for the collection and recycling of SEW within a conceptual framework that builds on the literature on environmental upgrading in global value chains (GVC). We advance the GVC literature by incorporating complementary insights from social anthropology and life-cycle assessments, which will enable end-user behaviour and waste disposal to be integrated into the GVC framework. It is shown how the conceptual framework could be applied in Kenya as an example of a regional hot spot for solar e-waste. We suggest three separate research areas that would ideally be combined as a coherent and programmatic approach to ensure the provision of solid inputs informing national SEW policies and systems of regulation.

1. Introduction

A central feature of the ongoing transition towards cleaner sources of energy is the rapidly increasing diffusion of solar photovoltaic (PV) technologies across both developed and developing countries. Global sales of small-scale, off-grid solar devices (OGS), including PV-based solar lanterns, solar chargers and solar home systems, reached 130 million between 2010 and 2017 and are expected to increase to up to 250 million units in 2017–2022 (GOGLA 2018). The sale of OGS devices is concentrated geographically in developing countries located in Sub-Saharan Africa (SSA), especially in East Africa, and to a lesser extent in South Asia and Latin America. OGS has thus become a key element in meeting the development imperative of ensuring affordable and sustainable sources of energy for all, which is high on the agenda in the Global South.

The rapid and significant diffusion of OGS is accompanied by an increasing number of private firms involved in producing, selling and distributing OGS to customers on a commercial basis in order to serve their electricity and lighting needs (Rolffs, Ockwell, and Byrne 2015). This market-based model of the provision of basic energy services to the off-grid population is widely considered a success in terms of green technology innovation, private-sector development and the replacement of carbon-intensive sources of energy, such as kerosene and diesel (Ondraczek 2013).

OGS suppliers are able to take advantage of the substantial improvements in the price and efficiency of core technology components (such as solar panels), the emergence of smart metering technologies and the widespread use of mobile phones and mobile payment schemes (Nygaard, Hansen, and Larsen 2016). These technological innovations have provided the basis for the emergence of so-called pay-as-you-go (PAYG) systems, a new business model adopted by OGS suppliers in order to target low-income customers in developing countries (Rolffs, Ockwell, and Byrne 2015). The PAYG model involves the installation of OGS with smart metering systems, which enables customers to purchase smaller amounts of electricity based on their needs by using mobile phones and payment systems, thereby avoiding high upfront costs.

Given the increasing sales of OGS, the solar electronic waste (SEW) being generated and disposed of is increasing at a similar pace, which has raised concerns about the overall environmental sustainability of the OGS industry. Such concerns have not only been raised by donors and solar industry representatives, but also by national governments who are worried about the environmental impacts and by key OGS suppliers who are under increasing reputational pressure (Murray and Corbyn 2018). Since environmental regulation and enforcement mechanisms are typically lacking in most developing countries, research aimed at identifying appropriate initiatives to effectively manage the collection and recycling of SEW merit attention. Except for a paper by Cross and Murray (2018), who tracked the breakdown, repair and disposal of OGS in Kenya, limited research has been carried out to understand the severity of the SEW challenge, including perspectives on how to address the challenge and the policy implications. Our aim in this paper is therefore to provide an overview of the current challenges of SEW in the Global South and to develop the building blocks of a conceptual framework for assessing the challenges and opportunities for the collection and recycling of SEW.

The paper is structured as follows. In Section 2 we provide an overview of the nature and scale of the emerging SEW problem. Section 3 describes the main elements of the suggested framework for assessing SEW disposal, while in Section 4 we provide an illustrative example of how the framework can be applied. Finally, our conclusions are provided in Section 5.

2. Off-grid solar e-waste

2.1. Small-scale off-grid solar devices

According to the Global Off-Grid Lighting Association (GOGLA 2016), small-scale OGS sold in the Global South can be grouped into two main product types based on the level of electricity that they provide:

(i) Pico solar devices with a capacity of between 0 and 11 Wp. These products consist of LED lamps and simple multi-light systems powered by a solar panel and can be used, for example, for lighting and mobile charging.

(ii) Solar home systems (SHS) with a capacity above 11 Wp and up to 200 Wp. These systems usually consist of a solar panel, a battery and multi-lighting systems, and can power several appliances such as TVs, fans, freezers and rice cookers.

Over the past decade, a number of cost-efficiency improvements in the core components of OGS have occurred not least in the areas of solar panels and LED lamps. The resulting improvements in the price/performance ratios of these systems has been a key factor in accelerating their diffusion (see Figure 1). Especially in rural areas in the Global South without access to the electricity grid, sales of 'packaged' OGS coupled to various electronic appliances, such as mobile-phone chargers, radios, televisions and refrigerators, have increased significantly. As such, OGS have become a key source of electrification for the off-grid population in rural areas alongside conventional sources, such as kerosene lamps, small generator sets and car batteries (Scott and Miller 2016).

Since 2010, when OGS started to spread worldwide, the market has grown steadily, reaching cumulative sales of around 130 million OGS devices in 2017 (GOGLA 2018). In 2016, 26 million OGS were sold globally, out of which 2.1 million was sold in Kenya (Cross and Murray 2018). Accordingly, Kenya may be referred to as a regional 'hot spot' where market uptake has progressed the most and where the generation of SEW is likely to reach the highest volumes. The existing projections illustrated in Figure 2 show that the market for OGS is expected to grow in the coming years, with annual sales almost tripling between 2018 and 2022, reaching an accumulated sales total of between 336 and 434 million by 2022 (GOGLA 2018). The potential global market is estimated at about 400 million OGS, which may become saturated within the next five to ten years. Given an expected lifetime of around four years for these devices, about 100 million OGS will continue to be sold and disposed of every year, in order to maintain existing services once market saturation is reached.

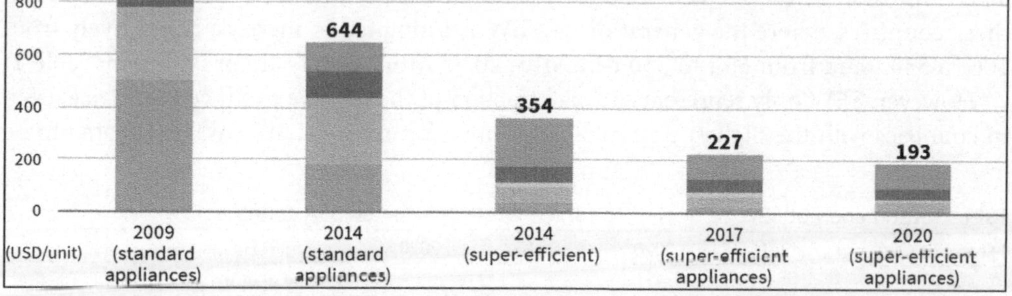

Figure 1. Estimated annual and accumulated sales of OGS devices globally (Millions).
Source: Modified from GOGLA (2018).

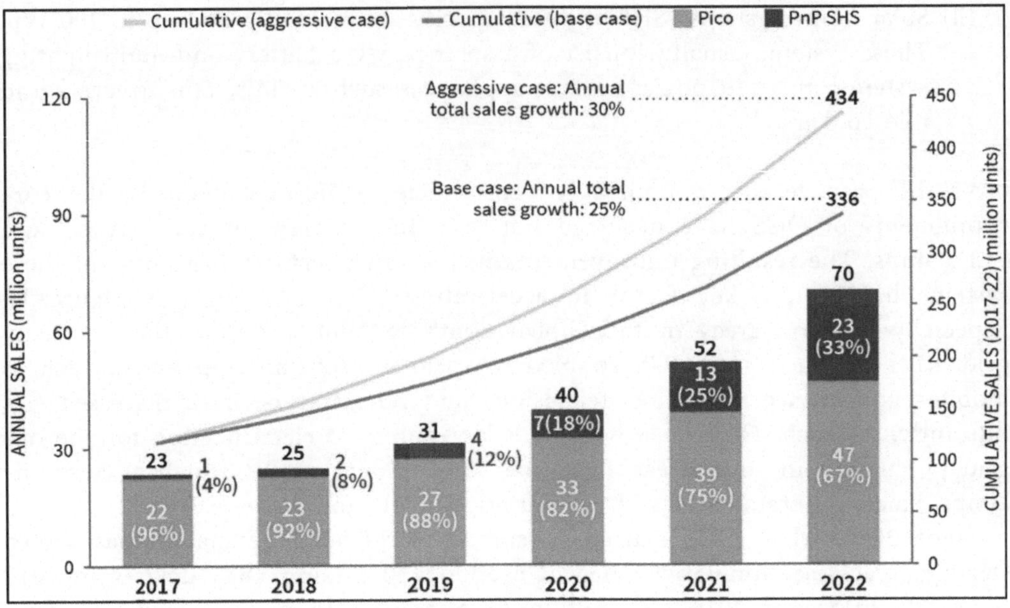

Figure 2. Estimated annual and accumulated sales of OGS devices globally (Millions).
Source: Modified from GOGLA (2018).

2.2. Projections of SEW

Bensch, Peters, and Sievert (2017) refer to the increased sales and diffusion of OGS in developing countries as representing an 'emerging disposal problem' because of the expected increase in generated e-waste. Indeed, GOGLA (2018) estimated that around 26 million OGS devices went out of use in 2017. It is the short lifetime of these devices that justifies the concerns around the amount of SEW they will generate. Recent country-level studies in Africa have produced estimates of the scale of the problem. Thus, based on data from GOGLA (2016), Magalini et al. (2016) calculated that in 2016 a total of 6700 t of OGS devices would be sold on the market in fourteen countries in SSA, which they estimated to have generated a total of 2200 t of SEW. In Kenya, around 700 t of SEW were discarded in 2016, a figure that is expected to reach 3,800 t in 2020. In spite of much lower diffusion rates of OGS in Nigeria and Rwanda, a similar trend is discernible in these countries, where the generation of SEW is estimated to increase respectively from 100 to 530 t and from 65 t to 350 t in 2016–2020. More details are provided in Table 1.

However, SEW only represents a small fraction of the e-waste produced in Africa, even in countries with the highest unit sales. In Kenya, for example, SEW was estimated to be

Table 1. Estimation of SEW compared to total e-waste in selected SSA countries.

Year	2016	2020	2016	2020	2017	(2016/2017)
T/year	OGS		SEW		Total e-waste	Off grid fraction
Kenya	1,500	5,700	700	3,800	55,000	1.3%
Nigeria	217	780	100	530	280,000	0.0%
Rwanda	144	520	65	350	9,742	0.7%
14 SSA countries	6,700	26,000	2,200	12,000	600,000	0.4%

Source: Compiled based on Magalini et al. (2016).

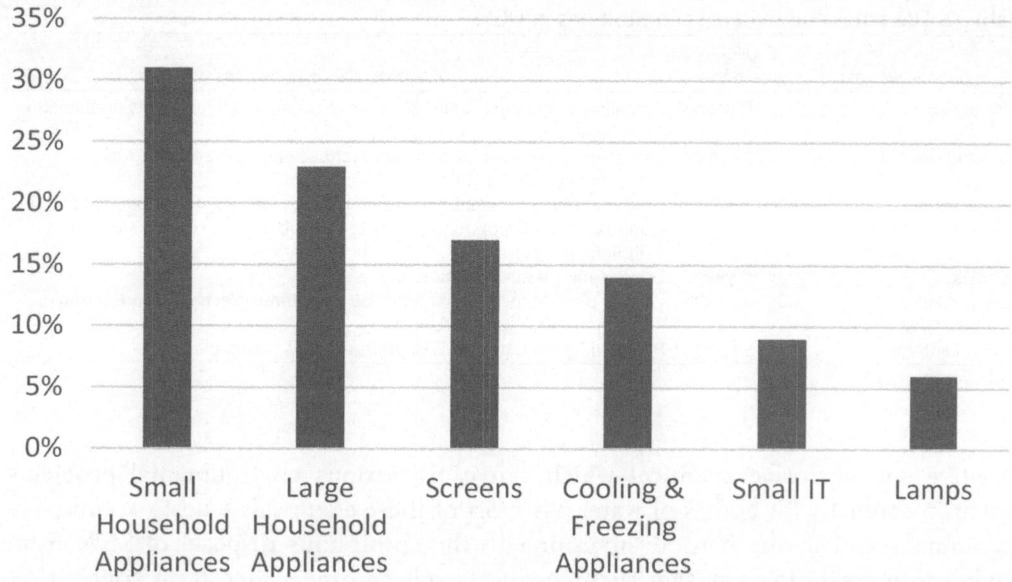

Figure 3. Estimated shares of the six main e-waste categories.
Source: Based on Magalini, Sinha-Khetriwal, and Munyambu (2017).

about 3% of the total e-waste generated in 2017, which is about half of the share of lamps (light bulbs). The share of all six main e-waste streams is shown in Figure 3.[1] Compared to global averages, the amount of e-waste generated in Africa is still relatively small (Mihai et al. 2019). The amounts of e-waste generated per capita in Europe and the Americas are estimated at around 16.6 kg/person and 11.6 kg/person respectively, while Asia produces 4.2 kg/person and Africa 1.9 kg/person (Corbyn, Martinez, and Cooke 2019a).

2.3. Environmental impacts

OGS contain various hazardous materials, such as lead, cadmium, mercury and sulphuric acid, which may cause serious adverse effects to humans and the environment (see Table 2). Lead is classified as a cumulative toxicant, harmful to the environment and to humans, its adverse effects mainly connected to the nervous system, blood pressure and kidneys. Lead is known to cause problems in many parts of the world, the highest burden being in developing countries, and its diffusion into the environment is also a result of the informal recycling of batteries (WHO 2010a). Mercury is a chemical that is highly toxic to human health, being mainly transported through the air and through deposits in bodies of water. Its adverse effects include damage to the nervous, digestive and immune systems, as well as lungs and kidneys (WHO 2007). Cadmium is a carcinogenic chemical, exposure to which is linked to bone damage and to lung and respiratory diseases (WHO 2010b; Latunussa et al. 2016). Such impacts can often be traced to the incineration of e-waste and the consequent inhaling of harmful chemicals (GIZ 2018). Sulphuric acid is a very

[1]For further details on the six main e-wastes streams, see e.g. Baldé et al. (2017).

Table 2. Life-times and material compositions of OGS.

Component groups	Expected life-times	Typical material compositions
PV panels	> 10 years	Crystalline silicon, glass, aluminium, copper and trace elements (indium, tin, and gallium).
Control devices	5 – 15 years	Printed circuit boards, solder paste, various electrical and electronic components and plastics.
Batteries	2 – 6 years	Lead-acid batteries: Lead, lead-oxide, plastics, electrolyte (sulphuric acid) Li-ion batteries: Graphite, various organic substances, copper, aluminium, lithium and plastics.
Cables	> 10 years	Copper and plastic insulation.
Equipment (lamps, radios, fans, TVs)	2 – 10 years	Various plastic types, aluminium, copper, various electrical and electronic components (e.g. microchips).
Solar Lanterns	3–5 years	PV panel, Li-ion battery, LEDs, printed circuit board, plastics.

Source: GIZ (2018).

reactive and corrosive chemical which can cause serious environmental problems when in contact with bodies of water. As most of these chemicals tend to accumulate in humans and in the natural environment, the continuous disposal of SEW from OGS can increase the risk that such chemicals will become a long-term concern.

2.4. Institutional and regulatory framework

In most SSA countries the infrastructure for the collection, handling and recycling of e-waste is poor and inadequate. As a result, most e-waste is disposed of in landfills, placed in temporary storage or deposited in nature or around people's houses (Magalini et al. 2016; Mihai et al. 2019). Due to this lack of regulation and proper systems for the collection and treatment of SEW and e-waste in general, the current and future handling and treatment of SEW and e-waste in SSA remains a huge challenge. Indeed, effective e-waste management systems are generally complicated by the often informal and uncoordinated structure of waste systems in most developing countries (van Welie et al. 2018). To this comes the highly decentralised nature of OGS product distribution channels across the countries, which may simply render collection and recycling schemes difficult to manage effectively and cost-efficiently (GIZ 2018).

Table 3. Status for e-waste and SEW regulation in selected developing countries.

Country	E-waste legislation	Availability of recycling infrastructure	OGS products specifically in scope	Batteries in scope	Main electrical and electronic equipment
Burundi	First draft	Fair/Poor	No	No	TV, Radio, Fans, water pumps
Kenya	Draft, pending final approval	Fair/Good	Potentially	Yes	TV, Radio, Fans, water pumps
Rwanda	Published	Fair/Good	Under discussion	Yes	TV, Radio, Fans, water pumps
Uganda	No draft	Poor	No	No	TV, Radio, Fans, water pumps
Ghana	Published, not enforced	Fair/Good	Yes/Partially	Yes	TV, Radio, Fans, water pumps
India	Published, not enforced for OGS	Fair/Good	Not in scope	Separate batteries legislation	TV, Refrigerators, AC

Source: Corbyn, Martinez, and Cooke (2019a).

Although e-waste regulation is currently being introduced in a number of SSA countries, as shown in Table 3 below, still mostly in draft form, and it generally remains unenforced. Furthermore, and most importantly, in most cases SEW does not fall within the scope of e-waste regulation.

2.5. Voluntary initiatives

To address these challenges, the Global Off-Grid Lighting Association (GOGLA) has taken some important initial steps. In GOGLA, attention has mainly been directed towards the financing of various forms of regulation and business models with the aim of identifying suitable economic incentive structures for e-waste management across the sector (Corbyn et al. 2019b). This perspective seems to be a logical starting point in order to address the specific actors that should bear the costs of managing SEW in the value chain. The well-known and longstanding guiding principle in e-waste regulation used around the world, referred to as 'extended producer responsibility' (OECD 2019), has provided the main source of inspiration in such deliberations. This principle entails that the main producers and suppliers of equipment should be held responsible for the waste and environmental impacts caused by the disposal of their products (McDonald and Pearce 2010). Accordingly, equipment suppliers should make a financial contribution to cover the costs of the collection and recycling of products at the end-of-life stage on either a voluntary or a mandatory basis. In 2014, GOGLA members, who include the main lead firms in the sector, adopted a voluntary agreement committing members to the principle of extended producer responsibility, under which they should:

- Develop products that can be easily maintained and repaired. Spare parts need to be made available
- Strategies to implement proper take-back and recycling should be envisaged in countries of operation
- Identify synergies in the use of standard resources and materials to facilitate separation during recycling and reuse
- Avoid the use of hazardous substances and find alternatives for them, if technically possible. If this is not possible, incentives for collection of the parts containing these hazardous substances should be developed

It could be argued that voluntary systems of this sort, which have been adopted in the absence of national regulation, indicate a genuine commitment on the part of the lead firms in the OGS industry to include e-waste management in their operations. However, the system does not include a reporting structure on performance, and since the system is not a mandatory requirement, no penalty system has been adopted to deal with cases of non-compliance. The voluntary system is thus one option to address the challenge of SEW, but there is an urgent need for research-based knowledge on the environmental and economic effectiveness of such systems, as well as for solid research proposals for their design, regulation and enforcement.

3. Towards a framework for assessing SEW disposal

In the following, we present the main conceptual building blocks that could serve as guidance for research on SEW going forward. The literature on global value chains is suggested as an overall basis for such research, as it provides a useful overarching framework for analysis. This perspective enables us to trace the flow of OGS materials along the OGS value chain, including the involvement of various actors and activities. This suggestion builds upon previous research undertaken by the authors in several research projects on the development of solar PV value chains in East Africa, focusing specifically on Kenya, Uganda and Tanzania (Hansen, Pedersen, and Nygaard 2015; Hansen 2018; Lema et al. 2018; Dal Maso et al. 2019; Bhamidipati, Hansen, and Haselip 2019).

3.1. The global value chain perspective

The literature on global value chains (GVC) provides a perspective for analysing the full range of activities along a value chain involved in bringing a product or service from its initial conception and production to its end use and beyond (Kaplinsky and Morris 2003). As trade in goods increasingly takes place in value chains, the GVC perspective has achieved prominence when it comes to understanding how the interlinked nature of production, trade and consumption evolve globally in specific sectors. Specifically, the GVC literature focuses on the role of the dominant actors within a given industry, the so-called 'lead firms', whose decisions and activities have repercussions throughout the entire value chain. Due to their central position in the value chain, lead firms are powerful actors that control the flows of information and resources, and the functional division of labour in the value chain (Dallas, Ponte, and Sturgeon 2019). For example, lead firms set the terms of chain membership, such as compliance with standards, the related incorporation or exclusion of other actors, and the re-allocation of value-adding activities (Nygaard and Bolwig 2018). A focus on the role of lead firms in relation to SEW seems to resonate well with the principle of extended producer responsibility referred to above.

Recent GVC literature has pointed out that chain-external actors, such as governments, standard-setters, multilateral institutions and NGOs, can significantly influence how lead firms operate (Ponte and Sturgeon 2014; Bolwig et al. 2010). In particular, the prevailing political and institutional framework conditions have been stressed as being among the key external factors influencing conditions in the value chain (Liu 2017). Thus, while GVC analyses involves a perspective focusing on the 'internal' conditions in the value chain governed by the lead firms, the external contextual conditions operating 'outside' value chains are equally important elements of such analyses. These institutional conditions may operate at various scales, ranging from the global to the national and the sub-national.

GVC research typically proceeds through a series of analytical steps starting with a mapping of the different segments of the value chain through which various materials and resources flow. This is combined with an analysis of the geographical location of these segments from the initial sourcing of input materials to the location of final consumption. Furthermore, specific attention is paid to analysing the role of the lead firms in governing activities in the value chain. Finally, the relevant aspects of the institutional context are analysed at the appropriate scales (Kaplinsky and Morris 2003).

GVC research is often undertaken in the context of South-North oriented value chains consisting of Northern buyers of intermediate goods procured from suppliers in the Global South. However, with the emergence of lead firms in the Global South and the growing levels of South-South trade, a more complex pattern of consumption and production in GVCs has evolved (Horner and Nadvi 2018). Indeed, under the new forms of polycentric trade, GVCs increasingly exhibit multi-polarity in both scope (global, domestic, regional) and orientation (North-South, South-North, South-South) (Ponte and Sturgeon 2014).

3.2. Environmental upgrading in GVCs

In the GVC literature the concept of 'upgrading' has been used to describe how the economic competitiveness of developing-country firms may be improved by shifting to more rewarding functional positions in a value chain or making products that have more value-added invested in them. According to Humphrey and Schmitz (2002), upgrading may occur by improving the quality of products, improving production efficiency or moving into high value-added activities in the chain, such as R&D.

Recently, the concept of 'environmental upgrading' has been introduced to improve the understanding of how the overall environmental performance of value chains can be improved and become more sustainable (Ponte, Gereffi, and Raj-Reichert 2019). According to Poulsen, Ponte, and Sornn-Friese (2018), the concept of 'environmental upgrading' denotes the process of improving the environmental impact of value chain operations, including production, processing, transport, consumption and waste disposal or recycling. Environmental upgrading may occur through the alteration of production systems that result in positive (or reduce negative) environmental outcomes (De Marchi, Maria, and Micelli 2013). In this context reduced environmental impacts may involve reducing carbon emissions, the depletion of natural resources, water and energy consumption, and after-use effects, such as waste, pollution and energy consumption (Khattak and Pinto 2018). The existing literature has generally ascribed importance to the role of the lead firms within the industry in pushing suppliers upstream in the value chain to improve their environmental performance (e.g. Khattak and Park 2018). Accordingly, De Marchi et al. (2019) suggest directing attention to analysing the specific external and internal drivers for lead firms to engage in environmental upgrading. Lead firms may, for example, respond to internal pressure from shareholders to increase economic benefits, such as by reducing energy consumption as part of their corporate social responsibility strategies or by improving the environmental quality of a product in order to reach new market segments. External drivers may include pressure on lead firms in the form of regulation and standards adopted globally or nationally, including voluntary sustainability standards, environmental regulation and pressure from industry associations and NGOs (Bush et al. 2015).

De Marchi et al. (2019) suggest that three main forms of environmental upgrading may occur in value chains related to improvements in processes, products or organisation (see Table 4). We suggest adding a fourth dimension to this understanding, namely 'product end-of-life improvements', which entail efforts by lead firms specifically to reduce end-of-life waste flows. Lead firms may, for example, introduce waste collection and recycling schemes for their products, including take-back schemes, initiatives to

Table 4. Types of environmental upgrading in GVCs.

Improvements	Characteristics of environmental upgrading
Process improvements	Entail eco-efficiency through the reorganisation of production systems or the use of superior technology (e.g. reduction of energy or materials used per unit of output);
Product improvements	Entail the development of sophisticated, environmentally-friendly product lines (e.g. usage of recyclable, recycled or natural inputs, de-materialisation of products, avoidance of use of toxic or impacting materials);
Organisational improvements	Entail organisational enhancement through a firm's overall way of doing business and managing the organisation, often related with the achievement of standards and certifications (such as LEED, or ISO14001).
Product end-of-life improvements	Entail reducing product end-of-life waste flows, for example, through waste collection and recycling schemes, including take-back schemes, initiatives to repair and refurbish used products, and the establishment of waste collection centres

Source: Modified based on De Marchi et al. (2019).

repair and refurbish used products, the establishment of waste collection centres, and efforts to extract valuable materials locally. This perspective, we argue, is particularly relevant in the context of value chains with end-markets in the Global South.

3.3. Incorporating end-user behaviour and waste into GVC research

While we consider the GVC perspective to be a useful basis for analysing solar e-waste disposal, we argue that in order to account fully for end-use behaviour and waste disposal in value chains, insights from complementary perspectives should be utilised (Bolwig et al. 2010). Specifically, we point to a need to integrate social anthropology to conduct household-level ethnographic research on solar product usage and disposal practices. Moreover, to address the economic and environmental potential for the collection and recycling of disposed OGS, we suggest incorporating insights from the literature on life-cycle assessment (LCA). Key insights from these bodies of literature will be elaborated in what follows.

3.3.1. The social anthropology of waste

Research in sociology and anthropology has long been concerned with analysing the complex relationship of how humans and physical artefacts interact. In the field of science and technology studies, for example, various strands of research have been devoted to the understanding of how society and technology co-evolve. Examples include studies of the social construction of technology (Bijker, Hughes, and Pinch 1987), the domestication of technology (Berker et al. 2006) and socio-technical systems (Ockwell et al. 2018). Studies within these fields of research typically revolve around analysing how the social practices and agency of individuals and groups of individuals emerge in relation to their engagement with technology (Enslev, Mirsal, and Winthereik 2018). The processes of sense-making, including how humans give meaning and value to material objects, are key elements of such analyses, as are structural issues, such as social norms, culture and ethnicity, class, gender and power. Studies of the waste sector that adopt an anthropological perspective have focused on waste as a particular form of materiality around which humans interact and construct their social lives (Drackner 2005; Moore 2012).

Intuitively, it would be logical to assume a direct and straightforward relationship between unit sales of products and waste disposal of products. However, a fundamental principle of anthropological research in the waste sector is the focus on understanding how social practices evolve in the course of converting goods and products into waste streams and how they reach their final sites of disposal (Liboiron 2014). This object of analyses in the literature on the social afterlife of products is referred to as the 'gap' between product breakdown and final disposal (Hetherington 2004). Research in this field thus involves understanding the entire range of activities and intermediaries involved in the repair, reuse, refurbishment, material extraction and repurposing of broken objects (Cross and Murray 2018). This typically involves in-depth and long-term fieldwork carried out among waste-pickers, waste-collectors and scrap-workers at landfill sites and open garbage dumps, as well as local repair shops, recyclers and scrap-dealers. Adopting a perspective that focuses on the social anthropology of waste may allow GVC analyses to include end-use behaviour by devoting attention to the process of how broken components move through homes, repair shops and recycling centres before reaching their final sites of disposal.

3.3.2. Life-cycle assessment of waste systems

Life-cycle assessment (LCA) is a widely used tool for assessing and analysing environmental impacts during the life-cycle of a product in order to compare the environmental performance of products providing the same function (Hauschild, Rosenbaum, and Olsen 2019). LCA thus provides a perspective with which to analyse a comprehensive range of environmental effects that are assignable to various products by quantifying the inputs and outputs of material flows and assessing how these flows affect the environment. Through its holistic perspective, LCA helps prevent burden-shifting (i.e. a shift in impact) both across impact categories and between life-cycle stages. As LCA is an ISO-standardised methodology, LCA studies undergo a series of well-defined analytical steps, including selecting the specific production system to be analysed, defining the boundaries of the system and identifying the specific impact categories to be analysed, such as toxicity, global warming and eutrophication (Horne, Grant, and Verghese 2019). Most LCA studies make use of large databases comprising various environmental-impact indicators across a range of product categories. LCA studies of the waste sector typically involve estimating the environmental impacts of end-of-life scenarios in order to identify the most environmentally efficient waste management option. Such scenarios may cover shorter or longer periods and include variation in the key assumptions, such as the specific impact categories assessed, the amount of waste generated and differences in waste-recycling rates. The resulting impacts may then be calculated as a basis for identifying the optimum scenarios. Scenarios may also include an assessment of the various economic impacts, including effects on local income generation and employment. These economic and social impacts may result from factoring in activities related to the collection, extraction, repair, recycling and refurbishment of broken products and related materials.

4. Integrated conceptual framework and empirical illustration

Based on the above, we suggest positioning future research on SEW in the context of chain-inspired analyses based on the GVC framework. In this case, we suggest undertaking research at the country-level as the overarching unit of analyses. Following De

Marchi et al. (2019), we argue in favour of focusing on specific 'reference points' along the value chain in order to assess the key actors and processes involved in environmental upgrading. Accordingly, we conceptualise five main nodes along the OGS value chain, three of which we suggest to include as the subjects of three distinct focus areas of research: (1) importing node (2) household node and (3) waste node (see Figure 4). As depicted in Figure 4, these three nodes involve the critical points along the value chain in which OGS devices are transformed, which represent relevant entry points in order to identify opportunities to change the flow of materials and avoid OGS devices turning into disposed SEW. The two remaining nodes in the value chain should form part of the analyses of the operations of the OGS lead firms positioned at the importing node in Figure 4 in relation to their OGS sourcing arrangements outside the focal country and their specific distribution channels within the focal country in question. This reflects the circumstance that lead firms in the OGS industry are typically based with headquarters within the focal countries in question and are mainly involved in the sale of imported equipment (such as solar panels) from suppliers abroad under their own brand names (Rolffs, Ockwell, and Byrne 2015).

While we consider the GVC perspective to be an overall framework of reference for analysing the structure and functioning of the entire value chain, including the role of the lead firms (node 1 in Figure 4), insights from social anthropology and LCA studies on waste should be brought in to analyse end-use behaviour in households (node 2 in Figure 4) and waste-recycling scenarios (node 3 in Figure 4). Accordingly, these complementary analytical perspectives becomes relevant to include in the analyses as the focus moves downstream in the value chain towards final disposal of SEW. However, we suggest devoting particular attention to the role of OGS lead firms in controlling the flow of OGS materials throughout the entire value chain from the import node to the final waste node, including their possible involvement in the collection and recycling of SEW.

In the following, we will describe how the conceptual framework described above could be used to guide future research on SEW. We will take as our point of departure an assessment of SEW in Kenya, which we treat as an example of a regional 'hot spot' in East

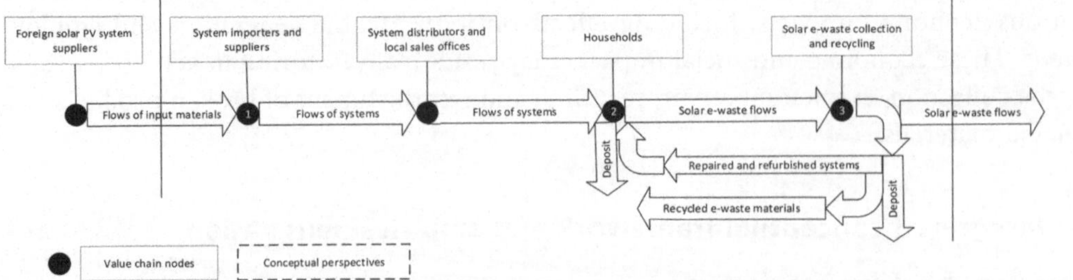

Figure 4. Integrated framework for assessment of SEW.
Source: Authors own elaboration.

Africa. Kenya has also been the primary focus of previous research carried out on SEW in Africa (see e.g. Cross and Murray 2018; Balasubramanian, Clare, and Ko 2019; Kumar and Turner 2020) and has featured prominently in the work conducted by the Global Off-Grid Lighting Association (GOGLA) on SEW (GOGLA 2019).

4.1. *Value chain analyses*

While the local assembly of panels and control units takes place in some developing countries, and while in some cases system components such as glass, aluminium and batteries are also sourced locally (Nygaard and Hansen 2016), typically OGS components are imported into developing countries by foreign suppliers (Few, Schmidt, and Gambhir 2019). The value chain for OGS products may therefore be treated as reaching their end-points in the importing countries. Therefore, improved understanding of the structure and dynamics of this local manufacturing and deployment chain is needed, rather than an analysis of the entire global OGS value chain. While the value chain depicted in Figure 4 is highly simplified, a detailed analysis of the OGS value chain in Kenya, for example, is likely to generate a more complex picture (Lema et al. 2018). Accordingly, whereas the node categories shown in Figure 4 would remain, the specific number and characteristics of the lead firms centrally positioned at the importing node should be analysed in depth, as should the specific nature of their distribution channels and forms of ownership. Such analyses, we argue, could pursue a conventional type of GVC analysis comprising the four main dimensions of relevance described in Section 3 above (i.e. mapping of value chain segments, geography, role of lead firms and institutional context). The research questions driving such analyses could be formulated as follows:

- How are value chains for OGS structured and governed?
- What are the institutional and regulatory frameworks for waste handling and recycling?
- To what extent are the leading suppliers of OGS currently involved in e-waste collection and recycling?
- What is the potential for increasing recycling by exercising environmental governance in the OGS value chain?

With a basis in the existing literature on environmental upgrading in GVCs, research aimed at addressing these questions should focus specifically on the role of lead firms in governing SEW in the value chain. Indeed, the practices and decisions of these lead firms are likely to have repercussions on SEW disposal throughout the entire value chain both nationally and locally. In this case, the lead firms are the companies responsible for importing, assembling and selling PAYG OGS as a package to local customers, which include companies such as BBOXX, Mobisol and Azuri Technologies (Rolffs, Ockwell, and Byrne 2015).

It appears that some of these lead firms have taken an active role in implementing initiatives aimed at reducing SEW. In Kenya, for example, the local battery-maker Associated Battery Manufacturers has stationed collectors in major towns to collect discarded lead batteries, which are recycled for lead.[2] Similarly, the Kenyan company Mobisol has established a network of local waste and battery recyclers in East Africa in order to collect and recycle solar electronic waste.[3] Some companies, such as BBOXX, have introduced replacement and take-back schemes and introduced repair centres in their local stores throughout the country (GOGLA 2019).

Some lead firms have thus taken action individually to improve the sustainability of their operations. However, it is not evident how many of these lead firms have implemented or are considering implementing initiatives aimed at improving SEW management as part of their business activities. Future research could therefore take its point of departure in the perspective provided by De Marchi et al. (2019), namely to focus specifically on analysing the internal and external drivers (and obstacles) for lead firms to adopt initiatives aimed at reducing SEW. In this case, it would appear that previous initiatives to address SEW by the industry's lead firms are part of proactive efforts aimed at avoiding reputational damage. Indeed, these companies' corporate social responsibility strategies should be considered in light of their heavy dependence on direct funding from foreign entities, which are subject to accountability constraints. As pointed out by Corbyn, Martinez, and Cooke (2019a;5) '*companies voluntarily establishing e-waste management operations and partnerships, even in the absence of regulation* [...] *are motivated by environmental and social protection, investor interests as well as brand and PR concerns*'. Research should analyse in further detail their underlying motivations and the resulting implications for waste disposal in the value chain.

4.2. End-use behaviour

While OGS end-users typically belong to lower income groups in the population, income levels, social status, levels of education, cultural backgrounds, geographical localisation and ethnicity may all vary. Such differences play a role in the choice of products purchased and the use patterns of individual households. Accordingly, household-level research is needed to improve our understanding of the reasons behind the choice of OGS purchased, including information on usage and motivation for final disposal. In particular, research could benefit from analysing whether and how specific components and materials of OGS are reused and recycled.

Indeed, as shown in a study undertaken by Cross and Murray (2018) in Kenya, a local network of recyclers and repair shops may form an effective local recycling economy by repairing items and extracting valuable resources before final disposal. Studying the social practices of household members as they interact with this local recycling economy could generate valuable insights into end-user behaviour (Ockwell et al. 2018). Indeed, adopting a social anthropological perspective would enable such research to trace the process of how materials move from rural homes to small-town repair clinics, where items are repaired, reused and repurposed, and to company workshops, where they are replaced, returned and stored. Following previous research in this field, particular attention could be paid to address how the perception of waste changes as it takes on new forms of materiality, value and meaning for those involved in the process (Drackner

[2]https://www.nation.co.ke/business/enterprise/Battery-maker-ABM-banks-on-recycling-to-face-importers/1954166-2062726-np7tkv/index.html.
[3]https://plugintheworld.com/press-releases/mobisol-tackles-recycling-challenges/.

2005). End-user practices and determinants of behavioural change in households could thus form part of such analyses.

Research in this area could also analyse how end-users respond to formalised systems of collecting and reusing SEW, including national (and sub-national) regulation and certification systems. Such analyses could be combined with assessments of the effectiveness of existing initiatives established at the national and municipality levels to collect and recycle SEW components and materials (see e.g. GIZ 2014). A focus on the value chain actors involved in the informal recycling economy merit attention in this regard as waste pickers may reject regulation aimed at formalising waste collection and recycling practices. The research questions driving the analyses to be pursued within the scope of the above could be formulated as follows:

- What are the practices of households in respect of the use, repair and disposal of waste from solar lanterns and solar home systems?
- Who are the specific actors involved in the process from product usage to waste deposit, including local collectors, repairmen, scrap-dealers and recyclers?
- What are the main determinants in changing existing practices in the direction of a higher degree of recycling?

4.3. Life-cycle assessment

According to Latunussa et al. (2016), LCA-based studies of the solar PV sector generally focus on the production process, and overlook the end-of-life stage. Moreover, previous LCA studies of the environmental impact of crystalline silicon-based solar PV have tended to concentrate on a highly aggregated level of the global solar PV industry (e.g. McDonald and Pearce 2010). While some LCA studies have adopted country-level analyses, such studies tend to focus on developed countries, which largely neglect the rapidly emerging solar PV markets in the Global South. Indeed, to our knowledge, systematic life-cycle assessments of the environmental impacts of solar PV in developing countries are entirely absent. Hence, as argued by Xu et al. (2018), LCA research is needed to analyse the final stage of the product cycle, which in concrete terms should effectively pursue analyses with the starting point at the final SEW disposal sites.

In the context of (solid) waste in developing countries, final disposal sites typically involve landfills and open garbage dumps (often associated with open waste burning). A key element of this research should focus on improving understanding of the impacts on both the natural environment and humans from the leakage of hazardous chemicals and materials from disposed OGS components. Following standard LCA methods, such research should aim at identifying the most environmentally feasible scenario among a range of scenarios with variation in the key assumptions (Horne, Grant, and Verghese). An assessment of the economic impacts of the different scenarios could be included as a basis for assessing the most economically feasible scenario in terms of local incomes and employment generation. Indeed, the sale of valuable materials extracted from discarded OGS, such as copper and aluminium, could potentially generate significant economic value for local communities (GIZ 2018). Dal Maso et al. (2019), for example, conducted a study of the overall sustainability of the solar PV mini grid sector in Kenya and pointed out the net effects of the environmental and economic impacts of an increase in the amount of SEW generated. Relevant research questions guiding such analyses include the following:

- Which SEW collection and recycling scenarios are most likely to emerge within the current economic and institutional context?
- Which scenarios offer the most environmentally and economically feasible pathways?
- What is the economic potential and feasibility of selling valuable materials extracted from SEW devices to export markets?

4.4. Discussion

While the analytical building blocks described above do not adhere to a fully developed and comprehensive conceptual framework, we believe that they do provide a useful direction to guide future research on SEW in developing countries. Such research could utilise our proposed framework as a starting point for empirical investigations and further theoretical refinement. The three suggested research areas could be undertaken in separation and in individual research projects. However, we argue that a coherent and programmatic approach comprising all three elements simultaneously would be ideal. Such an approach would allow findings emerging in one area of research to feed into ongoing research in the other areas. For example, as the understanding of the usage and disposal patterns of households emerges, such insights could feed directly into the development of likely scenarios to be included in the LCA research. Research applying the perspective presented here would generate new knowledge, which could feed into national SEW policies and systems of regulation.

5. Conclusion

We began this paper by alluding to the rapid and significant increase in the sale of small-scale, OGS devices in developing countries, especially in Sub-Saharan Africa. Indeed, a vibrant market has developed there, with numerous private companies selling OGS on a PAYG basis. OGS has thus come to form a central role in providing affordable sources of electricity to Africa's rural populations. This development is driven mainly by improvements in the price and efficiency of core components, the emergence of smart metering systems and the use of mobile phones and mobile payment schemes.

However, we also stressed the expected increase in the amount of e-waste generated as OGS devices enter their end-of-life stage, and pointed out the associated adverse environmental impacts for humans and other biological systems. If not properly addressed, SEW may become an example of the potentially dark side of innovation, in which the development of new technologies to solve specific problems, such as energy access, are typically accompanied by the creation of a set of new challenges. We argue that such negative side-effects of innovation should be addressed by conducting empirical research in order to identify effective mitigation measures.

Consequently, we have developed the building blocks of a conceptual framework for studying governance systems for the OGS life-cycle with the aim of identifying suitable SEW collection, handling and recycling strategies. We suggest situating this framework within the literature on environmental upgrading in global value chains. This perspective places emphasis on the role of lead firms that are positioned centrally at the importing node in the OGS value chain, which is well-aligned with the longstanding principle in the governance of e-waste management of extended producer responsibility. However, we also suggest incorporating complementary insights from social anthropology and life-cycle assessments in the waste sector in order to account for end-use behaviour and final waste disposal. Our contribution to the literature thus lies in the development of a framework that provides a basis for studying the final stages of the product cycle, which is typically neglected in GVC research. This perspective, we argue, is particularly appropriate in the context of the emergence of more complex and multipolar value chains with end-markets in the Global South. We have provided an illustrative example of how the perspective could be applied in the context of research on SEW in Kenya as an example of a regional SEW hot spot. We conclude by suggesting that, while the three suggested research areas could be undertaken as separate research projects, they should ideally be combined to provide a coherent and programmatic approach ensuring that solid inputs inform the development of national SEW policies and systems of regulation. As the sale of OGS devices has not yet reached its maximum, there is still time to provide such research-based inputs in order to avoid the off-grid solar market turning into a new environmental crisis.

Finally, the paper provides a number of relevant reflections to guide policy development in relation to improving the collection and recycling of SEW in the Global South. Firstly, the lead firms in the OGS industry are often greatly dependant on financial support from various national and international development organisations. Thus, incorporating requirements for SEW collection and recycling as part of their funding modalities could contribute substantially to incentivising lead firms to introduce product end-of-life improvements. Research could play a key role in raising the awareness in the development organisations providing funding to promote the OGS market of the challenges related to SEW. Secondly, as we have seen, national e-waste regulation is currently underway in a number of developing countries in SSA. However, as national regulation may often not be enforced effectively, there is a need for environmental NGOs to devote greater attention to the issue of SEW. Indeed, many environmental NGOs have tended to regard OGS as inherently environmentally friendly and have therefore not paid much attention to their potential harmful effects. In combination, efforts from environmental NGOs and development organisations could form a strong external pressure on lead firms. Thirdly, the GVC framework has been widely adopted by many government agencies in the developing world as an overarching perspective for managing and planning development programmes. The value chain-inspired perspective presented in this paper could thus function well as a guiding device to steer policy development on SEW management given its resonance with the current practices of national authorities. Finally, we believe that this paper provides relevant insights of broader relevance for the management of e-waste in the Global South in general. A value chain perspective, as suggested in this paper, could provide new source of inspiration on the central role of lead firms in governing the flow of materials throughout the entire value chain and the specific drivers for the lead firms to engage in environmental upgrading.

Disclosure statement

No potential conflict of interest was reported by the authors.

ORCID

Ulrich Elmer Hansen http://orcid.org/0000-0002-8997-6717
Ivan Nygaard http://orcid.org/0000-0002-7633-9241
Mirko Dal Maso http://orcid.org/0000-0002-2340-5911

References

Balasubramanian, S., D. Clare, and S. Ko, 2019. "Off-Grid Solar E-Waste: Impacts and Solutions in East Africa. Master's Thesis, Duke University." Available from: https://dukespace.lib.duke.edu/dspace/bitstream/handle/10161/18419/Balasubramanian_Clare_Ko-SolarEWaste.pdf?sequence=1&isAllowed=y (assessed 20 November 2019)

Baldé, C. P., V. Forti, V. Gray, R. Kuehr, and P. Stegmann, 2017. "The Global E-waste Monitor – 2017, United Nations University (UNU), International Telecommunication Union (ITU) & International Solid Waste Association (ISWA), Bonn/Geneva/Vienna."

Bensch, G., J. Peters, and M. Sievert. 2017. "The Lighting Transition in Rural Africa: From Kerosene to Battery-powered LED and the Emerging Disposal Problem." *Energy for Sustainable Development* 39: 13–20. doi:10.1016/j.esd.2017.03.004.

Berker, T., M. Hartmann, Y. Punie, and K. Ward. 2006. *Domestication of Media and Technology*. Open University Press:New York, USA.

Bhamidipati, P., U. Hansen, and J. Haselip. 2019. "Agency in Transition: The Role of Transnational Actors in the Development of the Off-grid Solar PV Regime in Uganda." *Environmental Innovation and Societal Transitions* 33: 30–44. doi:10.1016/j.eist.2019.02.001.

Bijker, W., T. Hughes, and T. Pinch, Eds.. 1987. *The Social Construction of Technological Systems: New Directions in the Sociology and History of Technology*. Cambridge, UK: MIT Press.

Bolwig, S., S. Ponte, A. Du Toit, L. Riisgaard, and N. Halberg. 2010. "Integrating Poverty and Environmental Concerns into Value-Chain Analysis: A Conceptual Framework." *Development Policy Review* 28 (2): 173–194. doi:10.1111/j.1467-7679.2010.00480.x.

Bush, S., P. Oosterveer, M. Bailey, and A. Mol. 2015. "Sustainability Governance of Chains and Networks: A Review and Future Outlook." *Journal of Cleaner Production* 107 (16): 8–19. doi:10.1016/j.jclepro.2014.10.019.

Corbyn, D., J. Martinez, and R. Cooke, 2019a. "E-waste Toolkit Module 4 Briefing Note: E-waste Regulation and Compliance." Available from: https://www.gogla.org/sites/default/files/resource_docs/gogla_e-waste-module-briefing-4_def.pdf (assessed November 20)

Corbyn, D., J. Martinez, R. Cooke, F. Magalini, and A. Kyriakopoulou, 2019b. "E-waste Toolkit Module 3 Briefing Note: The Financials of E-waste Management." Available from: https://www.gogla.org/sites/default/files/resource_docs/gogla_e-waste-module-briefing-3_def-web.pdf (assessed November 20)

Cross, J., and D. Murray. 2018. "The Afterlives of Solar Power: Waste and Repair off the Grid in Kenya." *Energy Research and Social Science* 44: 100–109. doi:10.1016/j.erss.2018.04.034.

Dal Maso, M., K. Olsen, Y. Dong, M. Pedersen, and M. Hauschild, 2019. "Sustainable Development Impacts of Nationally Determined Contributions: Assessing the Case of Mini-grids in Kenya, Climate Policy (Forthcoming)."

Dallas, M., S. Ponte, and T. Sturgeon. 2019. "Power in Global Value Chains." *Review of International Political Economy* 26 (4): 666–694. doi:10.1080/09692290.2019.1608284.

De Marchi, V., E. Maria, A. Krishnan, and S. Ponte. 2019. "Environmental Upgrading in Global Value Chains." In *Handbook on Global Value Chains*, edited by S. Ponte, G. Gereffi, and G. Raj-Reichert, 310–324. Cheltenham, UK: Edward Elgar.

De Marchi, V., E. Maria, and S. Micelli. 2013. "Environmental Strategies, Upgrading and Competitive Advantage in Global Value Chains." *Business Strategy and the Environment* 22 (1): 62–72. doi:10.1002/bse.1738.

Drackner, M. 2005. "What Is Waste? to Whom? an Anthropological Perspective on Garbage." *Waste Management and Research* 23 (3): 175–181. doi:10.1177/0734242X05054325.

Enslev, L., L. Mirsal, and B. Winthereik. 2018. "Anticipatory Infrastructural Practices: The Coming of Electricity in Rural Kenya." *Energy Research and Social Science* 44: 130–137. doi:10.1016/j.erss.2018.05.001.

Few, S., O. Schmidt, and A. Gambhir. 2019. "Energy Access through Electricity Storage: Insights from Technology Providers and Market Enablers." *Energy for Sustainable Development* 48: 1–10. doi:10.1016/j.esd.2018.09.008.

GIZ. 2014. *End of Life Solar Lamps Recycling Strategy Definition Kenya and Cameroon*. Deutsche Gesellschaft für Internationale Zusammenarbeit (GIZ) and Total Access to Solar (TATS). Assessed 20 April 2020. https://www.gogla.org/sites/default/files/recource_docs/recycling-study_final-report.pdf.

GIZ, 2018. "End-of-Life Management of Batteries in the Off-Grid Solar Sector, Deutsche Gesellschaft Für Internationale Zusammenarbeit (GIZ)."

GOGLA, 2016. "Off-Grid Solar Market Trends Report 2016." Bloomberg New Energy Finance, World Bank, IFC and Global Off-grid Lighting Association (GOGLA).

GOGLA, 2018. "Off-Grid Solar Market Trends Report 2018." Bloomberg New Energy Finance, World Bank, IFC and Global Off-grid Lighting Association (GOGLA)."

GOGLA, 2019. "GOGLA E-waste Festival Summary Report. The Global Off-Grid Lighting Association (GOGLA)." Available from: https://www.gogla.org/sites/default/files/summary_report_290819.pdf (assessed 20 November 2019)

Hansen, U., 2018. "The Insertion of Local Actors in the Global Value Chains for Solar PV and Wind Turbines in Kenya (IREK Working Paper No. 2)." Copenhagen/Nairobi/Eldoret: AAU, ACTS and MU. http://irekproject.net/files/2018/04/IREKPaper2.pdf

Hansen, U., M. Pedersen, and I. Nygaard. 2015. "Review of Solar PV Policies, Interventions and Diffusion in East Africa." *Renewable and Sustainable Energy Reviews* 46: 236–248. doi:10.1016/j.rser.2015.02.046.

Hauschild, M., R. Rosenbaum, and S. Olsen, Eds.. 2019. *Life Cycle Assessment: Theory and Practice*. Springer. https://doi.org/10.1007/978-3-319-56475-3

Hetherington, K. 2004. "Secondhandedness: Consumption, Disposal, and Absent Presence." *Environment and Planning D: Society and Space* 22 (1): 157–173. doi:10.1068/d315t.

Horne, R., T. Grant, and K. Verghese. 2019. *Life Cycle Assessment: Principles, Practice and Prospects*. Oxford: Csiro Publishing.

Horner, R., and K. Nadvi. 2018. "Global Value Chains and the Rise of the Global South: Unpacking Twenty-first Century Polycentric Trade." *Global Networks* 18 (2): 207–237. doi:10.1111/glob.12180.

Humphrey, J., and H. Schmitz. 2002. "How Does Insertion in Global Value Chains Affect Upgrading in Industrial Clusters?" *Regional Studies* 36 (9): 1017–1027. doi:10.1080/0034340022000022198.

Kaplinsky, R., and M. Morris. 2003. *A Handbook for Value Chain Research*. Institute of Development Studies, University of Sussex. Assessed 20 April 2020. http://www.value-chains.org/dyn/bds/docs/424/Value%20Chain%20Handbook%20Kaplinsky.pdf.

Khattak, A., and L. Pinto. 2018. "A Systematic Literature Review of the Environmental Upgrading in Global Value Chains and Future Research Agenda." *Journal of Distribution Science* 16 (11): 11–19. doi:10.15722/jds.16.11.201811.11.

Khattak, A., and Y. Park. 2018. "Environmental Upgrading of an Apparel Firm in Bangladesh: A Case Study of VIYELLATEX." *Emerald Emerging Markets Case Studies* 8 (3): 1–16. doi:10.1108/EEMCS-06-2017-0129.

Kumar, A., and B. Turner. 2020. "Sociomaterial Solar Waste: Afterlives and Lives after of Small Solar." In *Energy Justice Across Borders*, edited by G. Bombaerts, K. Jenkins, Y. A. Sanusi, and W. Guoyu, 155–173, Springer Open. https://doi.org/10.1007/978-3-030-24021-9.

Latunussa, C., F. Ardente, G. Blengini, and L. Mancini. 2016. "Life Cycle Assessment of an Innovative Recycling Process for Crystalline Silicon Photovoltaic Panels." *Solar Energy Materials and Solar Cells* 156: 101–111. doi:10.1016/j.solmat.2016.03.020.

Lema, R., R. Hanlin, U. Hansen, and C. Nzila. 2018. "Renewable Electrification and Local Capability Formation: Linkages and Interactive Learning." *Energy Policy* 117: 326–339. doi:10.1016/j.enpol.2018.02.011.

Liboiron, M., 2014. "Why Discard Studies? Discard Studies, Social Studies of Waste, Pollution and Externalities." Available from: https://discardstudies.com/2014/05/07/why-discard-studies /(assessed 20 November 2019)

Liu, Y. 2017. "The Dynamics of Local Upgrading in Globalizing Latecomer Regions: A Geographical Analysis." *Regional Studies* 51 (6): 880–893. doi:10.1080/00343404.2016.1143924.

Magalini, F., D. Sinha-Khetriwal, D. Rochat, J. Huismann, S. Munyambu, J. Oliech, I. Nnorom, and O. Mbera. 2016. *Electronic Waste (E-waste) Impacts and Mitigation Options in the Off-grid Renewable Energy Sector*. UK Department for International Development (DFID). Assessed 20 April 2020. https://assets.publishing.service.gov.uk/media/58482b3eed915d0b12000059/EoD_Report__20160825_E-Waste_Study_Final-31.08.16.pdf.

Magalini, F., D. Sinha-Khetriwal, and S. Munyambu. 2017. *Cost Benefit Analysis and Capacity Assessment for the Management of Electronic Waste (E-waste) in the Off-grid Renewable Energy Sector in Kenya*. UK Department for International Development (DFID). Accessed 20 April 2020. https://assets.publishing.service.gov.uk/media/5aec1a5ee5274a702130def2/Cost-Benefit_Analysis_and_Capacity_Assessment_for_the_Management_of_Electronic_Waste__E-Waste__in_the_Off-Grid_Renewable_Energy_Sector_in_Kenya.pdf.

McDonald, N., and J. Pearce. 2010. "Producer Responsibility and Recycling Solar Photovoltaic Modules." *Energy Policy* 38 (11): 7041–7047. doi:10.1016/j.enpol.2010.07.023.

Mihai, F., M. Gnoni, C. Meidiana, C. Ezeah, and V. Elia. 2019. *Waste Electrical and Electronic Equipment (WEEE): Flows, Quantities, and Management—A Global Scenario*. 1–34. Electronic Waste Management and Treatment Technology. Chapter 1. https://doi.org/10.1016/B978-0-12-816190-6.00001-7.

Moore, S. 2012. "Garbage Matters: Concepts in New Geographies of Waste." *Progress in Human Geography* 36 (6): 780–799. doi:10.1177/0309132512437077.

Murray, D., and D. Corbyn, 2018. "Industry Agrees It Is Time to Tackle Off-grid Solar Electronic Waste." Available from: https://www.gogla.org/about-us/blogs/industry-agrees-it-is-time-to-tackle-off-grid-solar-electronic-waste (assessed 20 November 2019)

Nygaard, I., and S. Bolwig. 2018. "The Rise and Fall of Foreign Private Investment in the Jatropha Biofuel Value Chain in Ghana." *Environmental Science and Policy* 84: 224–234. doi:10.1016/j.envsci.2017.08.007.

Nygaard, I., and U. Hansen, 2016. "Niche Development and Upgrading in the PV Value Chain: The Case of Local Assembly of PV Panels in Senegal." 2016 Annual Conference of the EU-SPRI Forum, Book of Abstracts, pp. 247–248. Assessed 20 April 2020. https://euspri-forum.eu/knowledge-base/book-of-abstract/.

Nygaard, I., U. E. Hansen, and T. Larsen, 2016. "The Emerging Market for Pico-scale Solar PV Systems in Sub-Saharan Africa: From Donor-supported Niches toward Market-based Rural Electrification." UNEP DTU Partnership.

Ockwell, D., R. Byrne, U. Hansen, J. Haselip, and I. Nygaard. 2018. "The Uptake and Diffusion of Solar Power in Africa: Socio-cultural and Political Insights on a Rapidly Emerging Socio-technical Transition." *Energy Research and Social Science* 44: 122–129. doi:10.1016/j.erss.2018.04.033.

OECD, 2019. "Extended Producer Responsibility." Organisation for Economic Co-operation and Development (OECD)." Available from: http://www.oecd.org/env/tools-evaluation/extendedproducerresponsibility.htm (assessed 20 November 2019)

Ondraczek, J. 2013. "The Sun Rises in the East (Of Africa): A Comparison of the Development and Status of Solar Energy Markets in Kenya and Tanzania'." *Energy Policy* 56: 407–417. doi:10.1016/j.enpol.2013.01.007.

Ponte, S., G. Gereffi, and G. Raj-Reichert, eds. 2019. *Handbook on Global Value Chains*. Cheltenham, UK: Edward Elgar.

Ponte, S., and T. Sturgeon. 2014. "Explaining Governance in Global Value Chains: A Modular Theory-building Effort." *Review of International Political Economy* 21 (1): 195–223. doi:10.1080/09692290.2013.809596.

Poulsen, R., S. Ponte, and H. Sornn-Friese. 2018. "Environmental Upgrading in Global Value Chains: The Potential and Limitations of Ports in the Greening of Maritime Transport." *Geoforum* 89: 83–95. doi:10.1016/j.geoforum.2018.01.011.

Rolffs, P., D. Ockwell, and R. Byrne. 2015. "Beyond Technology and Finance: Pay-as-you-go Sustainable Energy Access and Theories of Social Change." *Environment and Planning A: Economy and Space* 47 (12): 2609–2627. doi:10.1177/0308518X15615368.

Scott, A., and C. Miller. 2016. *Accelerating Access to Electricity in Africa with Off-grid Solar: The Market for Solar Household Solutions*. Overseas Development Institute (ODI). Assessed 20 April 2020. https://www.odi.org/sites/odi.org.uk/files/odi-assets/publications-opinion-files/10230.pdf.

van Welie, M., P. Cherunya, B. Truffer, and J. Murphy. 2018. "Analysing Transition Pathways in Developing Cities: The Case of Nairobi's Splintered Sanitation Regime." *Technological Forecasting and Social Change* 137: 259–271. doi:10.1016/j.techfore.2018.07.059.

WHO, 2007. "Exposure to Mercury: A Major Public Health Concern." World Health Organization (WHO). Available from: https://www.who.int/ipcs/features/mercury.pdf (assessed 20 November 2019)

WHO, 2010a. "Exposure to Lead: A Major Public Health Concern." World Health Organization (WHO). Available from: https://www.who.int/ipcs/features/lead.pdf (assessed 20 November 2019)

WHO, 2010b. "Exposure to Cadmium: A Major Public Health Concern." World Health Organization (WHO). Available from: https://www.who.int/ipcs/features/cadmium.pdf (assessed 20 November 2019)

Xu, Y., J. Li, Q. Tan, A. Peters, and C. Yang. 2018. "Global Status of Recycling Waste Solar Panels: A Review." *Waste Management* 75: 450–458. doi:10.1016/j.wasman.2018.01.036.

The dark side of the industrialisation of accountancy: innovation, commoditization, colonization and competitiveness

Emma C. Gardner and John R. Bryson

ABSTRACT

Technological innovation continues to play a fundamental role in disrupting many industries, but the impacts of digital innovation on accountancy and its effect on firms and individuals remain somewhat overlooked. Nevertheless, technological innovation is transforming accountancy work and business models as firms react to competition and regulatory restrictions. The analysis of sixty semi-structured interviews with UK accounting firms reveals how the adoption of new technology and alterations in regulations impacts on accounting practices. In turn, this research raises questions about the nature of professional occupations, the deskilling of accountancy and colonization of new service areas, challenging the extant conceptualisation of knowledge-intensive services.

1. Introduction

Accountancy firms and their constituent actors occupy an important role within our economies and as sites of embodied expertise labour (Bryson and Rusten 2005), safeguarding financial institutions and securing economic stability. The most fundamental role fulfilled by accountancy professionals, the auditing of company accounts safeguarding the interests of employees, shareholders and government, provided them with social closure (Weber (1922) 1978) and transformed accountancy into an archetypal profession. As such, this profession and its constituent members have a number of other attributes that qualify accountancy as a professional occupation including: acquisition of a systematic body of knowledge, the commitment to an ethical code, a shared culture and authority over non-(accountancy) professionals (Greenwood 1957). The satisfaction of these criteria provided the accountancy profession with security and economic rewards as a result of regulations enacted in the 1940s which excluded non-professionals from undertaking audit work (Hanlon 1994; Power 1997). However, accountancy is now experiencing new drivers of change which have presented, or continue to give rise to, interesting implications for accountancy professionals, and in turn the wider business ecosystem, and disrupt the roles traditionally enacted by accountants. This is part of an ongoing process of disruption and adaptation in response to technological and regulatory innovations combined with alterations in client behaviour and in the nature of competition.

The paper draws upon face-to-face interviews with accountants to identify and explore two interrelated drivers of disruptive change being experienced by accountants: technological innovations and changes in regulations. The dynamic interrelationships between these two drivers present an increasingly complex challenge for the accountancy industry and its professionals. While these factors are not novel, they are evolutionary, resulting in on-going firm and individual adaptations. Combined, these factors are part of a process of commoditization in which some accountancy services are increasingly selected on price rather than via reputation or third-party referrals. Innovation is altering the market structure of the accountancy industry as well as the everyday practices of accountancy. This paper's contribution is to explore the impacts technological innovation and regulatory changes are having on a profession. This research seeks to understand current regulatory and technological drivers of change and the impacts of these innovations on accountancy firms and individuals, and to identify adaptation strategies and their impacts. Some of these impacts highlight that there is perhaps always a darker side to innovation; innovation simultaneously destroys and creates tasks/jobs. These impacts also include deskilling combined with a reorientation of large accountancy practices from audit to non-audit work, which reflects a process of deprofessionalisation as accountancy practices focus on the provision of high-value services rather than regulated and more price-sensitive audit contracts.

This paper proceeds as follows. First, the research background outlines the contemporary accountancy industry and recent regulatory initiatives, as well as the extant knowledge on (de)professionalisation and technology adoption in accounting. Second, the research design and linked methodology that was developed to explore the impacts of technological and regulatory change on accountancy is presented. Third, the paper analyses how the role of technology is affecting accounting practices and the performance of accounting. Fourth, this paper concludes by summarising the impacts of technological and regulatory innovations on the accounting profession, highlighting areas for future research.

2. Research background

Accountancy is a business and professional service (BPS) with professionals employed by clients to provide advice and expertise. Some of this input is strategic and specialist and some relatively ubiquitous and generic but customised to meet the needs of a particular client. Some BPS sectors are professions whereby the ability to practice is based on the acquisition of systematic knowledge that is formally accredited by a professional body (Greenwood 1957). There are other important characteristics of being a professional; professionals have more control over the labour process, especially in terms of the rights and obligations expected for continued membership of their professional body (Freidson 1972: 56). Some BPS activities have not developed into fully recognised professional activities, for example marketing, public relations and management consultancy. Professional bodies exist in these areas, but non-members are still able to practice without sanction. Some client service inputs that are delivered by BPS firms are mandatory, for example the requirement for companies to have their accounts audited, while others are discretionary, for example employing a marketing consultant or a trained industrial designer. The diversity in BPS highlights the role played by quasi professions (Hearn 1982) and demonstrates the ambiguous boundary of professions in contemporary practice (Wilensky 1964). This has also led to the classification of traditional and new professions, exemplified by accountancy and management consulting respectively (Faulconbridge 2015).

The concept of competition is fundamental to neoliberal economic systems, but within professional services there are two contrasting views regarding the state of the field, particularly when considering the audit market. The last century saw a series of mergers and acquisitions between the largest accountancy firms, resulting in the emergence of several 'Big' practices. While there was a Big Six, PLCs arguably had a reasonable selection from which to choose when appointing an auditor; however, this choice was restricted with the merger of Price Waterhouse with Coopers and Lybrand in 1998 (ICAEW 2019), and was further compromised with the collapse of Arthur Andersen in the wake of the Enron scandal in 2002.[1] This left four 'Big' companies: Deloitte, EY, KPMG and PwC. Although there are a plethora of other firms with audit licences, such as RSM or Grant Thornton, they do not operate on the same global scale as the so-called Big 4 (see e.g. Financial Reporting Council 2017: 45) and do not have a comparable depth of resources with which to provide services to multinational clients. Thus, while in theory, there are more than four providers in the market for audit services, in reality, there are limited options for larger blue chip companies (Crump 2013), some of whom are path dependant in their choice of service provider giving rise to decades-long audit firm appointments (see e.g. Wallace 2015). It could therefore be argued that the audit market is not competitive at all, a view which is held by regulators.

Innovations including developments in computer programming over the past three decades have presented significant opportunities for accountancy practices. Whereas beforehand, the embodied expertise ascribed with knowledge-intensive services and the professions was just that, the development of computer programmes and software packages has provided a new means for knowledge codification and dissemination. Thus, it increased the transferability of knowledge, by extracting knowledge (or know-how) from individuals (Abbott 1988) and facilitating its application to new contexts by other individuals (Suddaby and Greenwood 2001). Regulation is synonymous with professionalisation (Greenwood 1957), determining that those who wish to practice accounting must acquire a systematic body of knowledge and attain the relevant qualifications. Despite various layers of governance and regulation, the accountancy industry, however, has regularly been criticised for unethical conduct. The most infamous case surrounds the activities of Arthur Andersen, auditors and consultants to the former failed energy giant, Enron, whose dual role evidently compromised the integrity of their audit opinion (Chicago Tribune 2002). This marked a turning point, with regulators introducing the Sarbanes-Oxley act in the US. Similar regulations were developed and introduced in the UK with the intention to remove any conflict of interests between the provision of audit services and consulting services (ICAEW 2019bb). Nevertheless, accountancy firms reacted by innovating to circumnavigate regulatory constraints, which resulted in a fresh wave of concerns over accounting activities due to continued

[1] http://news.bbc.co.uk/1/hi/business/2,047,122.stm.

questionable conduct by auditors, who were often simultaneously pursing more lucrative non-audit contracts from their audit clientele. Thus, 2014 saw the ratification of mandatory auditor rotation in the European Union, which was implemented in 2016, forcing those companies that must statutorily procure an external audit, such as Public Interest Entities (PIEs), to change provider at least every twenty years, with a requirement to put their audit work out to tender once per decade (ICAEW 2019c). This strategy was designed to increase auditor independence, reduce auditor-client embeddedness and promote competition within the industry. New regulations also limited the amount of non-audit work that accountancy practices could conduct for an existing audit client, capping this revenue at 70% of the audit fee (Ibid.).

Accountancy is a highly regulated profession, but accountancy firms provide a range of professionalised or regulated and unregulated services to clients. This is an important point. Regulatory change, combined with innovation, encourages accountancy firms to alter the balance between the provision of regulated and unregulated services. This has been labelled as a process of 'colonization' by Suddaby and Greenwood (2001) in which large accountancy firms migrate into adjacent professional jurisdictions, for example law, cybersecurity and technology, in response to enhanced competition and commodification. Nevertheless, it is worth noting that our research has identified that BPS colonization predominantly involves the provision of non-professionalised rather than professionalised services.

There are push and pull drivers behind this migration or colonization. On the one hand, client problems are often complex, requiring accountancy firms to develop new areas of expertise. This is an ongoing process in which expertise providers must alter the balance of provided services. On the other hand, BPS firms are engaged in a continual process of innovation as they attempt to differentiate themselves from competitors. There are two drivers behind this innovation. First, BPS firms develop new services, but clients rapidly internalise these services reducing demand for external consultancy-related inputs, facilitated by technological offerings. Second, competitors develop their own versions of these new services. There is thus a continual race to innovate. Innovation is restricted in the provision of heavily regulated professional expertise which is another driver behind colonization into unregulated practice areas.

Colonization can be conceptualised as a process by which accountancy firms deprofessionalise. This process of deprofessionalisation operates at two levels. First, colonization alters the balance within accountancy firms towards the provision of unregulated and unprofessionalised services. Second, the application of technology to the provision of audit and other regulated accountancy services has altered everyday accountancy practices. Automation has altered task content as part of a displacement effect in which capital substitutes for tasks previously performed by labour. For white-collar workers in accounting, sales, logistics, trading and some managerial occupations, some tasks have been replaced by specialist software and artificial intelligence (Acemoglu and Restrepo 2019: 4).

The application of technology to audit by clients and accountancy firms is part of a process in which traditional auditing skills are experiencing a process of deskilling combined with reskilling (Bravermann 1974). This reflects the impacts of the application of technology to BPS firms and the emergence of *protech* or the professional services version of fintech or proptech. Deskilling and the degradation of work is not usually associated with professionalised labour, but it does occur (Suddaby and Muzio 2015). The ongoing application of technology, combined with commoditization, to audit and accountancy has shifted the balance in the delivery of accountancy services towards more routine, fragmented and deskilled tasks. The affected functions typically make use of knowledge management systems and databases, which enable firms to efficiently deal with providing solutions to standardised client problems (Faulconbridge 2015). Part of this process involving the degradation of audit and accountancy work is reflected in the expectations placed on accountants that they must work longer hours. It is also reflected in the shift to provide non-audit services. The status of accountancy, audit and accountants is altering. At the same time, there has been a decline in professionally qualified accounting academics being employed in the accounting academic community (Paisey and Paisey 2017).

The process of regulatory and technological evolution is not finite; innovation continues to present new challenges for the industry which requires that social scientists continue to explore the impacts innovation is having on what is arguably one of the archetypal professions. Given this, it is important to understand how firms are responding and adapting to these types of innovations by exploring both firm level and actor adaptions.

3. Methodology

To explore and understand the impact and responses to regulatory and technological innovations, we adopted a comparative case study research design using a qualitative methodology that included sixty semi-structured intensive interviews with accounting professionals working in 16 firms. These informants were all working within the West Midlands (UK) and were chosen via purposeful sampling to represent different service lines and hierarchical levels, to ensure that the sample could provide a holistic insight into the changing nature of the profession and of accountancy practices. In this way, we selected an intentionally heterogeneous sample, to determine whether our findings would be applicable across contexts (Mason 2002; Robinson 2014). Despite difficulties in obtaining access, reflecting the time pressures experienced by accountants, the number of interviews overcomes concerns relating to selection bias.

Our interpretive approach enabled us to explore the profession from the perspective of its constituent actors, but by making use of additional case material, we are able to attest to the validity of our findings. In addition, we took steps to ensure the trustworthiness of our data by speaking with multiple informants from each firm where possible, and by incorporating 16 firms in our sample, we avoid prioritising novelty (Tokatli 2015). We also avoided hindsight bias by focusing our interviews on the current practices that existed in the firm and on the everyday practices experienced by each interviewee (Aguinis and Solarino 2019).

Our interview guide was based on themes from extant academic and practitioner literature and was piloted on representatives from Recognised Supervisory Bodies. The interviews took place in 2014 and 2015 and were audio-recorded where permission was granted, before being transcribed. As the audit rotation policy had been announced in 2014, to be implemented in 2016 (European Commission 2014), the interviews enabled us to gain insight into firm-level adaptions in preparation for this change. In all cases, detailed notes were taken during the interviews. The interview material was then analysed using Nvivo, with data coded based upon predetermined themes, but also in relation to emergent themes within the transcripts. We then undertook axial coding to identify linkages between the themes (Allen 2017).

We uncovered new realities about the nature of accounting work as a result of the application of new technology and present two primary effects of technological and regulatory drivers of change; the rebalancing of firm activities between audit and non-audit work and an ongoing process that centres on the deprofessionalisation of accountancy.

4. Analysis: exploring accountancy adaptations and the performance of accounting

The interviews confirmed that the heavily regulated nature of audit and accountancy has led to somewhat standardised outputs. This challenges accountancy firms to innovate their value propositions in order to attract and retain clients in the non-statutory audit market, as well as to justify fees in the PIE arena to cost-conscious clients. One respondent noted that '*Your client will always try and push your fee down in audit*' (Medium Firm #2).

As part of this product innovation, firms are transforming what they physically deliver to the client and '*… typically the bigger the business, the bigger the report used to be. You would get volumes of the thing. You could use them as a weapon to hurt people; they were seriously humungous documents. We basically took things back to what are the absolutely key decision-making things that somebody needs to know*' (Big 4 Firm #1). This reflects both a focus on strategic inputs, but also reduces the time and cost of providing service inputs to clients. Firm are also incorporating technical advances including analytics into their core offerings and one respondent noted that '*we deliver analytics as standard as part of the audit*' (Big 4 Firm #2). In this way, they are adding an advisory element to the audit work.

The incorporation of technology is not improving the quality of audit work; a sample of 325 Big 4 audits was reviewed by the Financial Reporting Council between 2013/14-2016/17, with 83 found to be requiring improvements (Table 1). Importantly, 'good' audits also include those requiring limited improvements; thus, this category still raises questions about audit quality as the balance between those requiring no and limited improvements is unreported. Moreover, process innovations are taking place as firms adopt technological innovations reconfiguring the ways that firm routines are operationalised. For example, '*the cloud has been really facilitative in allowing alternative delivery models*' (Big 4 Firm #2), however, over time, some of these technological innovations,

Table 1. Big 4 audit quality pre and post regulatory change.

Year	Audits reviewed	Audits classed as good	Audits requiring improvement
2013–14	68	45	23
2014–15	78	54	24
2015–16	89	70	19
2016–17	90	73	17

Financial Reporting Council (2017b, 2017c, 2017d, 2017e).

such as software packages, become available on the open market, enabling clients undertake tasks themselves in-house.

Accountancy firms have long sought to diversify their service offerings. Sometimes this occurs within traditional 'accountancy' boundaries, such as a Medium Firm #4 which positioned itself to advise on pension auto-enrolment by hiring a pension specialist. Firms were also diversifying into new industries including cybersecurity: '*Technology is used a lot in all our functions. There's a lot of particularly in the tech advisory space, so that's cyber security, big data, things like that … clients now have just got so much information that is electronic, that they need guidance around storage, analysing …* ' (Big 4 Firm #3). The friction that this creates within the profession is acknowledged, although previous regulatory iterations have not curtailed the conflicts that this colonization can create.

There is also the risk that mandatory audit rotation for large client firms will not ameliorate audit quality: '*if you change auditor every ten years, there is a danger that in the first couple of years of a new audit firm there will be a dip in quality, just because it takes you a long time to get your head round something*' (Big 4 Firm #2). To prepare for the revenue threats that would come with audit rotation policy, accountancy firms began to intensify their colonization of adjacent professional jurisdictions. Firms began to colonize new areas with one respondent noting that '*We're now a law firm as well. We can provide legal services, so we're definitely more multi-disciplined*' (Big 4 Firm #3). More often, however, they were redoubling their efforts in colonizing industries in which they already had a foothold, such as management consulting and other forms of advisory. This offensive strategy was being operationalised primarily through acquisitions; PwC acquired Booz & Co, KPMG acquired High Point Rendel, EY acquired Seren and Integrc, and Deloitte acquired Kaisen Consulting, all of which took place as part of the multitude of deals to acquire consultancy firms that were agreed by the Big 4 in 2014–15 (Table 2).

This flurry of colonization activity affects the balance of accountancy versus non-accounting, or audit versus non-audit work being carried out within firms. The latter is more profitable for firms, with prices for these service products determined by '*what the market can bear*' (Big 4 Firm #2). The shifting balance in fee income between these competing domains is evident in Tables 3 and 4.

Table 2. Number of Big 4 consulting deals.

	Deloitte	EY	KPMG	PwC
Deals in 2014	14	11	18	13
Deals in 2015	19	3	18	11
Total	33	14	36	24

Consultancy.uk (2017)

Table 3. The balance between audit and non-audit income, 2015 (£m).

Firm	Audit Income	Tax Income	Advisory Income
Deloitte	706	562	622
EY	550	470	559
KPMG	495	381	598
PwC	1025	714	495
Grant Thornton	134	92	Unknown

Accountancy Age (2015)

Table 4. The profitability of audit (2016 year end).

Firm	Number of PIEs audited	Audit Income (£m)	Income from Non-Audit Work to Audit Clients (£m)	Non-Audit Client Fee Income (£m)
Deloitte	338	430	212	2064
EY	215	395	275	1480
KPMG	539	488	243	1337
PwC	325	659	384	1905

Financial Reporting Council (2017)

The introduction of audit rotation regulation was intended to protect the sanctity of audit within firms and increase auditor independence. Nevertheless, the shift towards revenues generated from other jurisdictions highlights the decline in the importance of audit work for the Big 4 (Table 4). The Big 4 are transforming into suppliers of expertise that is not directly linked to the provision of accountancy services and is not linked to the provision of non-audit services to companies whose accounts they are auditing. The Big 4 are shifting away from being accountancy firms to providers of specialist non-audit services to clients.

The changing role of audit has, however, brought with it a need for new skills. The increased deployment of technology in the audit process has resulted in a shift in the types of technical skills required by accountancy firms. Algorithms and machine learning have been incorporated into accountancy by both clients and accountants. Accountancy is becoming more of a technical process based around the application of algorithms to increase operational efficiency rather than being predominantly based on the delivery of professional accountancy inputs. This has worrying consequences for the skill base of the profession. One respondent noted that:

> '*… in some respects, it'll be fantastic that we've got all these new whizzy tools and things that can happen. But, the flip-side of it is that our ability to train people to then operate effectively at a senior level, when they haven't done any of the low-level auditing and really, really understood the nuts and bolts of what we do, because it has been done by computer, will then make it harder for them to then exercise judgement*' (Big 4 Firm #2).

The danger is that accountants will no longer have the training and practice required to understand audit processes and their related problems and challenges in any depth. The accountant's expertise will be partly hidden within computer code and machine learning. This exposes the accountant, clients and wider society to errors that are hidden within code and bias that emerge within machine learning-based systems. The auditor's role is changing to become the face of the firm and the role is shifting towards that of being a client relationship manager tasked with securing non-audit revenue from audit clients or attempting to transfer audit clients into more lucrative non-audit clients, post-rotation of an audit contact. Thus, '*if we lose an audit client, it may actually benefit us to end that; they may still want to use us in an advisory capacity*' (Big 4 Firm #3). Deskilling of audit work is occurring, as is a reskilling of the accountant, to cope with new market conditions. This is also a more general phenomenon across accountancy firms, as more individuals working within colonized areas are being enveloped into practices, altering the core skills base of these firms with the balance between professional and non-professional activities shifting towards the latter.

The shift in the evolving identity of the accountant from doer to overseer and client manager is altering the expectations placed upon professional workers. Technological advances have enabled the traditional boundaries of work to be eroded, providing constant connectivity between managers and employees, and clients and professionals, elongating the working day. Technological innovation including smartphones, combined with the shift in emphasis to client relationship management, means that for many accountants the boundaries between work and everyday living become increasingly blurred. One accountant described a working day as follows '*I normally get home at 7pm [. . .] Once I've caught up on the day, eaten and whatever, I will then get my laptop out and do another couple of hours work*' (Big 4 Firm #2). This is exacerbated by the rotation policy, which increases workloads as firms must bid for existing and new contracts without increasing audit staffing levels. This in turn compromises the work-life balance of accountants and has implications for their health and wellbeing.

The evolution of the profession, and the changing role of the accountant, could also explain the turnover of staff within firms; sixteen of the interviewees had left their firms to work in-house in the private sector since data collection. While some of this turnover could be attributed to natural attrition or chance factors, the interviews signalled low morale and stress within the industry. One respondent stated that '*I don't sleep massively well, I probably get five hours a night on weekdays which is not the most healthy and I understand that it's because I don't switch off*' (Big 4 Firm #3). This inability, or discouragement from 'switching off' indicates that work life balance concerns could be a key factor behind career moves. The challenge is that the accountant has limited free time and thus '*if an email comes in Saturday and it's something urgent, I'd rather know about it and deal with it than ignore it until Monday and have someone be annoyed at you*' (Big 4 Firm #3).

5. Discussion: colonization combined with deskilling

Our analysis has identified that there are two processes at work within the Big 4 accountancy firms. On the one side, there are alternations in the balance between the delivery of professional services and other types of non-audit services. On the other side, there are alterations in the delivery of accountancy services including deskilling linked to the application of software solutions by clients and accountancy firms. These two processes reflect a reworking of the everyday practices involved in the delivery of professionalised and other types of expertise to clients.

Any meaningful definition of professionalism within BPS must reflect the realities of everyday practices including the mismatch between professional codes and practice. To Hoyle and Wallace (2007) this reflects the 'irony of presentation' in which the image of a profession 'is not wholly congruent with the reality of its daily practices' (2007: 19). This has led Evans (2008) to distinguish between three different states of professionalism. First, *demanded* or *requested* professionalism reflecting specific professional service level demands. Second, *prescribed* professionalism reflected in recommended professional service levels, and, third, professionalism that is *enacted*. To Evans (2008: 13) it is this third state of professionalism that reflects the reality of everyday practice. This conceptual framework has been developed from an analysis of professionalism in the provision of educational services. Nevertheless, this approach provides an important conceptual framework that can be applied to exploring accountancy and other

BPS industries. The Big 4 accountancy firms' everyday practices are changing in response to alterations in regulations and innovations, but these changes are in advance of any attempt to regulate new practices. A linked ecology (Abbott 2005) has developed in which the Big 4 accountancy firms are adapting to alterations in regulations combined with innovations and these adaptations are challenging the jurisdictional boundaries of accountancy firms and the accountancy profession itself. One difficulty is that *enacted* professionalism is always one step ahead of *demanded* or *prescribed* professionalism. This lag explains difficulties that emerge with the quality of professional inputs as enacted professionalism challenges existing professional conventions.

This paper has explored the impact of regulatory and technological innovations on the accountancy industry focusing on understanding the implications of these drivers of change on organisations and professionals. In doing this, it has uncovered how regulatory and technological innovations are resulting in three levels of innovation within organisations. First, it has identified process innovations, enabled through the application of new technologies to enhance existing routines within firms. These include altering the balance between audit and non-audit work in response to rotation regulations; regulation restricts non-audit fee income from audit clients, and then accountancy firms prioritise growing other service lines over audit. Secondly, it has identified product innovations, whereby technological innovations are utilised to differentiate existing outputs, for example by combining analytics with the core audit offering. Thirdly, it has identified business model innovations as accountancy firms expanded their 'professional' boundaries by colonizing adjacent service areas to diversify their portfolio of services provided and perhaps, more importantly, to alter the balance between audit and non-audit service delivery.

One of the paper's key contributions is to identify that accountancy firms' colonization is into non-professionalised rather than professionalised service lines. Social closure (Weber (1922)1978: 638) is one of the defining characteristics of a profession by which one group closes off opportunities to another group of outsiders (Murphy 1988). Social closure enabled providers of professional services to control and restrict entry, but it also inhibits innovation and differentiation by product and process; professional bodies restrict and regulate innovation. Further research is required into the relationship between professionalisation and the regulation of process and product innovation and the ways in which the everyday enactment of professional practices occurs in advance of demanded or prescribed professionalism.

For professional service lines, differentiation has conventionally been based on third party referrals and reputations. Ongoing colonization of non-professionalised services increases the ability of accountancy firms to compete on product and process innovations, as these non-professionalised services are market-driven and can avoid or resist commoditization through continual innovation. The new audit rotation regulation has had a perverse consequence. It was intended to increase auditor independence and enhance the quality of client audit, but it has also encouraged accountancy firms to focus their attention on the provision of non-audit income lines. This is consistent with the assertion of Hansnata (2016), that competition within the accountancy industry is undermining the independence of audit.

The responses to audit rotation and fee limits on non-audit work have resulted in a variety of organisational adaptations. This is a new enacted professionality – a new iteration of the accounting profession – as routines are adapted to service audit clients but also to ensure, maintain and enhance profitability. This includes the envelopment of non-professional activities into the audit function, such as analytics and commercial/sales capabilities. Nevertheless, the dynamics of these inputs, and the resultant effects on the professional service product of the audit, are not yet regulated. In this way, the Big firms, which colonize new areas of expertise by incorporating them into their professional jurisdictions, are outrunning their professional regulators. Thus, there is a time lag between enacted professionalism and the types of professionalism prescribed to firms (Evans 2008: 12). This is evidenced by the time it takes regulators to respond with initiatives intended to combat audit quality concerns.

KPMG was singled out by the Financial Reporting Council in 2018 for a significant deterioration in the quality of its audit work, supported by its role as auditor to the now-bankrupt Carillion (Davies 2018). In response, it ceased offering non-audit services to its audit clients in November 2018 (Crisp 2018); a rather altruistic move given potential revenues such services generate, but perhaps one that was necessary to salvage or protect this firm's reputation. In January 2019, KPMG announced that it had decided not to re-bid for the audit contract with Aston Martin when the work was to go out to tender later that year, despite holding a strong position as the current incumbent provider (Skoulding 2019). Having provided this firm with audit services for over a decade, KPMG may benefit more from providing non-audit inputs to Aston Martin. This leads to the question of whether innovations in regulation are turning providers away from the audit market completely and whether the recent actions of KPMG will ripple throughout the industry. If so, this strategy will have important impacts on the market for audit services. More recently, in July 2019, PwC (2019), traditionally the market-leader in the PIE audit market, announced plans to invest £30 million annually in its audit arm. While this could be seen as a genuine exercise in improving audit quality, it is important to note that of the 26 of this firm's audits which were reviewed by the Financial Reporting Council (2019) during the past year, six were classed as requiring improvement. Moreover, it is important to understand the dynamics of PwC's proposed investment; while some of this funding will be channelled into its external audit function, it will also fund the creation of a new digital audit team intended to focus on the provision of internal audits, cyber security and risk reviews (Marriage 2019). This represents a shift in the types of audit services provided by this firm and is partly another form of colonization.

The different facets of innovation outlined in this research each have implications for individual workers, firms and the profession. Traditionally, employees within large accounting firms would have completed their Chartered qualifications, with those remaining with the firm post-qualification employed in the traditional accounting divisions of audit or taxation, or perhaps consulting. With the service diversification that such firms have pursued, however, audit and related accountancy services are becoming an increasingly peripheral element of their business models. As such, it is necessary to question the professional status these firms are afforded by society, when an increasing number of their employees neither holding nor practicing professional occupations (Greenwood 1957). This is further complicated by the fact that technology has evolved to replace many of the manual tasks that used to be undertaken by accountants; the assistive role of software is well established and the application of knowledge databases to BPS is acknowledged as a threat to embodied expertise (Faulconbridge 2015).

As firms seek to lower their cost-base, technology poses a threat to knowledge workers through the possibilities afforded by artificial intelligence. From a firm perspective, the productivity this would achieve is attractive, but it places professional workers in a precarious position: it is usually low-skilled, manual jobs that are considered to be at risk from automation, not middle-class, professional occupations (Autor 2015; Bryson 2018; Acemoglu and Restrepo 2019). If, or perhaps when, automation is further deployed within the accounting industry, the effect on the labour market will be interesting. One impact will be a further reduction in the quality of audit processes as critical thought in audit processes may be lost and replaced by automated processes. This is contrary to extant literature (Hansnata 2016) which asserts that automation improves audit quality. Technological innovation displaces control over expert knowledge away from professionals (Suddaby and Muzio 2015). In the short term, the deteriorating working conditions within the industry due to enhanced workload and client demands have important negative consequences for professional employees. The erosion of the boundary between home and workplacess is diminishing the possibility for an appropriate work–life balance, affecting the wellbeing of professionals as well as reducing the attractiveness of the profession to new entrants.

6. Conclusion

This paper has identified and explored two interrelated drivers of change being experienced by accountancy firms and their professional employees: technological innovation and regulation. A central theme of this paper and this special issue is to explore the darker sides of innovation. Innovation emerges in response to disruption including alterations in the regulatory environment. Disruption displaces existing markets, regulations, technologies and industries. Disruption is simultaneously destructive and creative. It is creative as firms, industries and individuals adapt and innovate and it is destructive as existing tasks, occupations, firms and industries are displaced and altered.

In the regulatory sphere, this research has identified some of the issues presented by the introduction of mandatory auditor rotation. Recently, there have been calls for new regulations to separate audit and consulting activities (Competition and Markets Authority 2018), seemingly recognising that the legalisation introduced in 2016 has failed to improve audit quality and ensure auditor independence. Yet, due to deeply enmeshed networks between actors in regulatory bodies and the larger accounting firms, it remains unlikely that truly effective regulation will be enacted to safeguard the sanctity of the audit profession, when such policies would be damaging to the Big 4.

This research has highlighted how despite perceived positive technological and regulatory innovations in the UK accounting industry, a closer analysis reveals a darker side to these innovations. Technological advances with their revolutionary possibilities have resulted in the redesign of jobs including the deskilling of previously expertise-laden embodied professional expertise. Embodied professional expertise is being substituted by algorithms and machine learning or a process of translation is occurring in which embodied expertise is being translated into computer code through an ongoing process of digitisation. The ongoing application of artificial intelligence and machine learning will require new regulatory supervision. The lag time between the enactment of innovations in accountancy and service practices and their capture and incorporation into requested and prescribed professional standards is a major problem. The speed at which enactment occurs makes it difficult for regulators to adjust, exposing clients and national economies to potential threats including failures in the quality of service delivery.

While this research presents a novel account of technological and regulatory innovations in the accounting industry, it is necessary to extend this analysis to other industries. This includes applying the conceptual distinction between *enacted, demanded* or *prescribed* professionalism to exploring the tensions between innovation, regulation and service quality in other BPS sectors and occupations. In this analysis, it is important to explore the adaptation strategies and the impacts on firms, employees, including work–life balance, as well as on clients and service quality. The commodification of knowledge is occurring in other BPS industries, such as management consulting and software design, whereby somewhat standardised solutions can be developed that can be slightly customised to clients' individual requirements. Colonization may also be occurring. Nevertheless, the colonization of new professional jurisdictions is in some ways distinct in accountancy. Accountancy is a professional service rather than just a business service and this professionalism is intended to play an important safeguarding role in the wider economy.

The legal profession, which operates a similar role, is similarly fragmented into high value and low-value areas of work whereby innovations in both areas are changing the dynamics of practice; this could provide an interesting setting in which to explore the impact of digital innovation on other professional firms and orientations. Nevertheless, the interpretive, deductive design adopted in this study of accountancy does not allow us to attest to the generalisability of our findings to other industries, and we seek to avoid decontextualised, abstract claims (Robinson 2014). Future research should seek to understand how the drivers of change within the accountancy industry which we have investigated are impacting upon adjacent professional fields, to lead to the generation of testable hypotheses. What is certain, however, is that technological and regulatory innovations have the potential to enable product and process innovations in all industries. Nevertheless, the primary challenge is that the positive and negative consequences, including the darker sides to innovation, of technological adoptions and firm/employee adaptations, must be explored before implementation. It is important that regulation, including the self-regulation of a profession, must keep pace with alterations in enacted professionalism.

Disclosure statement

No potential conflict of interest was reported by the authors.

ORCID

John R. Bryson http://orcid.org/0000-0002-6435-8402

References

Abbott, A. 1988. *The System of Professions: An Essay on the Division of Expert Labor.* Chicago: University of Chicago Press.

Abbott, A. 2005. "Linked Ecologies: States and Universities as Environments for Professions." *Sociological Theory* 23 (3): 245–274. doi:10.1111/j.0735-2751.2005.00253.x.

Accountancy Age. 2015. "Accountancy Age Top 50+50 Survey 2015." https://www.accountancyage.com/2015/08/24/accountancy-age-top-5050-survey-2015/

Acemoglu, D., and P. Restrepo. 2019. "Automation and New Tasks: How Technology Displaces and Reinstates Labor." *Journal of Economic Perspectives* 33 (2): 3–30. doi:10.1257/jep.33.2.3.

Aguinis, H., and A. M. Solarino. 2019. "Transparency and Replicability in Qualitative Research: The Case of Interviews with Elite Informants." *Strategic Management Journal* 40 (8): 1291–1315.

Allen, M. 2017. *The SAGE Encyclopedia of Communication Research Methods.* Thousand Oaks, CA: Sage.

Autor, D. H. 2015. "Why are There Still so Many Jobs? The History and Future of Workplace Automation." *Journal of Economic Perspectives* 29 (3): 3–30. doi:10.1257/jep.29.3.3.

Bravermann, H. 1974. *Labour and Monopoly Capial: The Degradtion of Work in the Twentieth Century.* New York: Monthly Review Press.

Bryson, J. R. 2018. "Divisions of Labour, Technology and the Transformation of Work: Worker to Robot or Self-Employment and the Gig Economy." In *Handbook on the Geographies of Regions and Territories*, edited by A. Paasi, J. Harrison, and M. Jones, 141–152. Cheltenham: Edward Elgar.

Bryson, J. R., and G. Rusten. 2005. "Spatial Divisions of Expertise: Knowledge Intensive Business Service Firms and Regional Development in Norway." *The Service Industries Journal* 25 (8): 959–977. doi:10.1080/02642060500237353.

Chicago Tribune. 2002. "The Fall of Andersen." http://www.chicagotribune.com/news/chi-0209010315sep01-story

Competition and Markets Authority. 2018. "CMA Proposes Reforms to Improve Competition in Audit Sector." https://www.gov.uk/government/news/cma-proposes-reforms-to-improve-competition-in-audit-sector

Consultancy.uk. 2017. "The 10 Largest Consulting Firms in the World." https://www.consultancy.uk/news/14018/the-10-largest-consulting-firms-in-the-world

Crisp, W. 2018. "KPMG to Stop Providing Extra Services for Audit Clients." *The Telegraph.* https://www.telegraph.co.uk/business/2018/11/08/kpmg-stop-providing-extra-services-audit-clients/

Crump, R. 2013. "Mid-Tier Handed "Permission to Play" in FTSE 350 Audit Market Shake Up." *Accountancy Age.* https://www.accountancyage.com/aa/analysis/2284500/midtier-handed-permission-to-play-in-ftse-350-audit-market

Davies, R. 2018. "KPMG Singled Out in Critical Report on Audit Industry." *The Guardian.* https://www.theguardian.com/business/2018/jun/18/kpmg-singled-out-in-critical-report-on-audit-industry

European Commission. 2014. "Directive 2014/56/EU of the European Parliament and of the Council of 16 April 2014 Amending Directive 2006/43/EC on Statutory Audits of Annual Accounts and Consolidated Accounts." https://eur-lex.europa.eu/legal-content/EN/TXT/?uri=celex%3A32014L0056

Evans, L. 2008. "Professionalism, Professionality and the Development of Education Professionals." *British Journal of Educational Studies* 56 (1): 20–38. doi:10.1111/j.1467-8527.2007.00392.x.

Faulconbridge, J. 2015. "Knowledge and Learning in Professional Service Firms." In *The Oxford Handbook of Professional Service Firms*, edited by L. Empson, D. Muzio, J. P. Broschak, and C. S. Hinings, 425–451. Oxford: Oxford University Press.

Financial Reporting Council. 2017. "Key Facts and Trends in the Accountancy Profession." https://www.frc.org.uk/getattachment/77fc8390-d0d1-4bfe-9938-8965ff72b1b2/Key-Facts-and-Trends-2017.pdf

Financial Reporting Council. 2017b. "Deloitte LLP Audit Quality Inspection." https://www.frc.org.uk/getattachment/b3e36efe-f328-47e6-8644-26987f3936fd/DELOITTE-LLP-Audit-Quality-Inspection-16-17.pdf

Financial Reporting Council. 2017c. "*Ernst* & Young LLP Audit Quality Inspection." https://www.frc.org.uk/getattachment/87fed5be-634e-43ed-8d66-824262dccd2d/EY-LLP-Audit-Quality-Inspection-16-17.pdf

Financial Reporting Council. 2017d. "KPMG LLP KPMG Audit PLC Audit Quality Inspection." https://www.frc.org.uk/getattachment/84251a1d-be78-4590-b284-ea47d6c8cc75/KPMG-LLP-Audit-Quality-Inspection-16-17.pdf

Financial Reporting Council. 2017e. "PwC LLP Audit Quality Inspection." https://www.frc.org.uk/getattachment/268c1302-ed75-4313-bed5-e78f5a9003e5/PwC-LLP-Audit-Quality-Inspection.pdf

Financial Reporting Council. 2019. "PwC LLP Audit Quality Inspection." https://www.frc.org.uk/getattachment/b2653cf4-9c44-4be2-bb87-d38cf5d3b44a/PwC-LLP-Public-Report-2018-19.pdf

Freidson, E. 1972. "Professionalization and the Organization of Middle-Class Labour in Postindustrial Society." *The Sociological Review* 20 (1): 47–59. doi:10.1111/j.1467-954X.1972.tb03209.x.

Greenwood, E. 1957. "Attributes of a Profession." *Social Work* 2 (3): 45–55.

Hanlon, G. 1994. *The Commercialisation of Accountancy*. Hampshire: Macmillan.

Hansnata, M. 2016. *The Impact of Digital Innovation on the Social Structure of Professional Public Accounting Practice in Australia*. https://pdfs.semanticscholar.org/3087/8389ab005c2a2664d916639ca8e370c1f3a8.pdf

Hearn, J. 1982. "Notes on Patriarcy, Professionalization and the Semi-Professions." *Sociology* 16 (2): 184–202. doi:10.1177/0038038582016002002.

Hoyle, E., and M. Wallace. 2007. "Educational Reform: An Ironic Perspective." *Educational Management, Administration & Leadership* 35 (1): 9–25. doi:10.1177/1741143207071383.

ICAEW. 2019. "Price Waterhouse." https://www.icaew.com/library/historical-resources/guide-to-historical-resources/firm-histories/whats-in-a-name/price-waterhouse

ICAEW. 2019b. "Companies (Audit, Investigations and Community Enterprise) Act 2004." https://www.icaew.com/en/technical/legal-and-regulatory/company-law/modernising-uk-company-law-history/companies-audit-investigations-and-community-enterprise-act-2004

ICAEW. 2019c. "Implementation of European Audit Reforms." https://www.icaew.com/en/technical/ethics/auditor-independence/implementation-of-european-audit-reforms

Marriage, M. 2019. "PwC to Split Audit Practice and Hire 500 Experienced Staff." https://www.ft.com/content/47f89a82-86e0-11e9-97ea-05ac2431f453

Mason, J. 2002. *Qualitative Researching*. London: Sage.

Murphy, R. 1988. *Social Closure: The Theory of Monopolization and Exclusion*. Oxford: Clarendon Press.

Paisey, C., and N. J. Paisey. 2017. "The Decline of the Professionally-Qualified Accounting Academic: Recruitment into the Accounting Academic Community." *Accounting Forum* 41 (2): 57–76. doi:10.1016/j.accfor.2017.02.001.

Power, M. 1997. *The Audit Society: Rituals of Verification*. Oxford: Oxford University Press.

PwC. 2019. "PwC Announces a New Action Plan to Strengthen Its Focus on Audit Quality." https://www.pwc.co.uk/press-room/press-releases/pwc-announces-a-new-action-plan-to-strengthen-its-focus-on-audit-quality.html

Robinson, O. C. 2014. "Sampling in Interview-Based Qualitative Research: A Theoretical and Practical Guide." *Qualitative Research in Psychology* 11 (1): 25–41. doi:10.1080/14780887.2013.801543.

Skoulding, L. 2019. "Aston Martin Searches for New Auditor as KPMG Declines Opportunity to Re-Pitch." Accountancy Age. https://www.accountancyage.com/2019/01/03/aston-martin-searches-for-new-auditor-as-kpmg-declines-opportunity-to-re-pitch/

Suddaby, R., and R. Greenwood. 2001. "Colonizing Knowledge: Commodification as a Dynamic of Jurisdictional Expansion in Professional Service Firms." *Human Relations* 54 (7): 933–953. doi:10.1177/0018726701547007.

Suddaby, R., and D. Muzio. 2015. "Theoretical Perspectives on the Professions." In *The Oxford Handbook of Professional Service Firms*, edited by L. Empson, D. Muzio, J. P. Broschak, and C. S. Hinings, 25–47. Oxford: Oxford University Press.

Tokatli, N. 2015. "Single-Firm Case Studies in Economic Geography: Some Methodological Reflections on the Case of Zara." *Journal of Economic Geography* 15: 631–647. doi:10.1093/jeg/lbu013.

Wallace, T. 2015. "Barclays Hires KPMG as Its Auditor after 120 Years with PwC." *The Telegraph.* http://www.telegraph.co.uk/finance/newsbysector/epic/barc/11716302/Barclays-hires-KPMG-as-its-auditor-after-120-years-with-PwC.html

Weber, M. (1922) 1978. *Economy and Society: An Outline of Interpretive Sociology.* Berkeley: University of California Press.

Wilensky, H. L. 1964. "The Professionalization of Everyone?" *American Journal of Sociology* 70 (2): 137–158. doi:10.1086/223790.

OPEN ACCESS

The grey zones of technological innovation: negative unintended consequences as a counterbalance to novelty

Maureen McKelvey and Rögnvaldur J. Saemundsson

ABSTRACT

The purpose of this article is to better understand the challenges of avoiding the dark side of technological innovation. Specifically, we analyse 10 public investigations started as a reaction to a major crisis in regenerative medicine at the Karolinska Institute, Sweden, associated with the clinician-scientist Paolo Macchiarini. We interpret the reaction as an attempt to restore the balance between the stimulation and regulation of technological innovation processes by clarifying ambiguities in the regulation at the interface between research and practice. We conceptualise these ambiguities as grey zones – situations when it is unclear if the benefits of experimentation outweigh its risks – and propose that grey zones are continually created and resolved as actors in innovation governance systems counterbalance the generation of novelty and the risk of negative unintended consequences.

1. Introduction

Technological innovation has both positive and negative impacts upon society (Stirling 2017). While the negative impacts are in this special issue referred to as the dark side of innovation one might (with implicit reference to the movie Star Wars) refer to the positive impacts as its light side. The new knowledge generated as outcomes of scientific, technological and innovation processes can lead to great benefits for society – such as increasing living standards through improved food safety or new medical procedures – but the process itself, which is inherently a process involving trial-and-error, may expose people, or the environment, to undue risks or even harm (Moreno 2001). The purpose of this article is to better understand the challenges of avoiding the dark side of technological innovation. We explore how actors in the innovation governance system counterbalance the stimulation of novelty for future benefits for society with the regulation of novelty, in order to avoid the risk of unintended consequences.

Let us first consider why this issue matters, when we consider the relationship between public policy and innovation governance. The rationale for public policy makers to spend money to stimulate science, technology and innovation is the expectation of future positive effects, such as improving living standards or solving societal challenges (Mazzucato 2018). In this positive view, society benefits from science and technology in the long-run through the introduction of new products like the iPhone, new services like microloans, new clinical practice to cure malaria, and new organisational forms to bring ideas to market through entrepreneurship. Specifically, the type of public policy that we are interested in here relates to science, technology and innovation, often known as innovation policy (Borrás and Edquist 2019). Following previous research, innovation policy is here defined as public policy intended to stimulate the creation and use of new scientific and technical knowledge, due to its potential future benefits for society, by promoting collaboration among multiple actors involved in developing science, technology and innovation, such as universities, research institutes, industrial firms, and user communities (Edler and Fagerberg 2017). Contributions also analyse how collaboration occurs, when the development of science, technology and innovation relies upon diverse types of actors, which each have their own aims and incentives for participation (McKelvey, Zaring, and Szücs 2019). The broader concept of innovation governance is used to denote how these different types of actors, together with public regulatory authorities, shape how innovations are produced, introduced, and diffused by collectively regulating issues of societal concern (Borrás and Edler 2014). As the stimulation of science, technology, and innovation are issues of societal concern, especially for medical innovation (McKelvey, Saemundsson, and Zaring 2018), we explicitly include it as a part of innovation governance.

We have chosen to study a major crisis within medicine, more specifically in regenerative medicine in Sweden. The crisis, associated with the clinician-scientist Paolo Macchiarini, played out at the prestigious Karolinska Institute – the home to the Nobel Prize in Medicine – and sparked a public outcry due to the belief that the involved clinician-scientists had overstepped ethical boundaries in their search for novel treatments. We consider the crisis as a focusing device that we use to analyse the dynamics of innovation governance more generally. Crises within science, technology and innovation tend to lead to public concerns about the negative effects and associated risks of innovation, which in turn promote additional demands for more inclusive, responsible and transparent innovation processes (Stilgoe, Owen, and MacNaghten 2013). Hence, the empirical study enables us to analyse the deliberations and reactions of actors in an innovation governance system that find themselves needing to respond to negative unintended consequences of a strong stimulation for research excellence and major industrial impact. We propose an extension of the concept of grey zones – which we define as situations when it is unclear if the benefits of experimentation outweigh its risks. We also propose that grey zones are continually created and resolved as innovation governance systems counterbalance the generation of novelty and the risk of negative unintended consequences.

2. Conceptual framework

This section discusses and defines relevant concepts and relationships for the conceptual framework guiding our study, which is based on three components, namely: a conceptualisation of innovation governance; an evolutionary perspective on the generation and use of new scientific and technical knowledge; and the empirical setting of medical innovation. At the end, we visualise our framework.

Our first component is a definition of innovation governance. From a broad interdisciplinary perspective, Borrás and Edler (2014) define innovation governance as how actors involved in the development of science, technology and innovation collectively regulate issues of societal concern by shaping how innovations are produced, introduced, and diffused in society. Thus, one starting point here is that the innovation governance system has two purposes, namely to stimulate and to regulate innovation processes.

Innovation researchers have long held that innovation is an uncertain and complex process because it relies upon mutual interactions and knowledge flows across private actors like firms, societal actors like universities, non-profit organisations and professional societies, as well as public actors like government agencies (Fagerberg, Mowery, and Nelson 2005). From this stream of literature, we extract the broad notion that actors involved in innovation governance attempt to both stimulate and regulate the generation and use of scientific and technical knowledge useful for innovation. We also note that innovation governance is complex, in the sense that a wide variety of actors are involved, each with their different aims and different ideas of which issues are of societal concern. Furthermore, the uncertainty about the outcomes of innovation processes – especially if they are related to emerging science and technology – require some forms of tentative governance that is revised as new knowledge becomes available, e.g. about negative unintended consequences (Kuhlmann, Stegmaier, and Konrad 2019).

Therefore, in this paper, we define innovation governance as a dynamic, systemic, multi-level process involving diverse sets of actors – which may be directly engaged in, or external to, the innovation process – and which are concerned with the stimulation and regulation of innovation. This builds upon McKelvey, Zaring, and Szücs (2019), which articulates the collective action and public resource pools as well as monitoring problems in the development of science and technology involving multiple actors. Analysing the institutions and monitoring aspects of the innovation governance system is important, because both the public sector and the private sector are investing money into research and development for science and technology. The public sector wants new knowledge which can be diffused widely and improve society, whereas the private sector is primarily concerned with developing and using such knowledge to generate profits through innovations. This perspective has been applied to medical research and innovation specifically. McKelvey, Saemundsson, and Zaring (2018) define in more detail what constitutes collective action at the intersection between medical research and clinical practice. They argue collective action for medical research is particularly difficult to regulate because it usually takes place in collaboration between universities, university hospitals, and industrial firms. Moreover, they propose a distinction between self-regulation by the actors involved in medical research and clinical practice as compared to external regulation by government agencies and similar.

Therefore, we follow a definition from McKelvey, Zaring, and Szücs (2019) and also used for medical innovation in McKelvey, Saemundsson, and Zaring (2018). Here, we conceptualise that collective action and public resource pools are created jointly by public and private actors. Furthermore, these diverse actors develop norms, incentives and institutions, to support the interaction required for the collective action to be successful and which also helps monitoring and regulating undesirable behaviour not conforming to the common interests of the actors involved.

Summarising our first component, we define innovation governance as attempts by diverse set of actors to stimulate and regulate the generation and use of new knowledge involved in science, technology and innovation processes. Thus, we see innovation governance as involving two separate goals, namely on the one hand stimulating science, technology and innovation as collective action due to the expected future benefits and, on the other hand, the regulation of the collective action process to avoid undesirable behaviour. Furthermore, we acknowledge the tentative and dynamic nature of innovation governance, especially in the context of emerging science and technology.

Our second component is an evolutionary theoretical perspective applied to the generation and use of new scientific and technical knowledge. At the most general level, an evolutionary perspective for social sciences is concerned with how the characteristics of entities evolve as the entities adapt to their environment, and usually involves a framework specifying the generation of variety, selection processes and the retention of elements selected (Campbell 1987; Nelson and Winter 1982). In relation to our topic of scientific and technical knowledge useful for innovation, we draw upon a stream of literature that has connected the evolutionary process of variety generation and selective retention to the process of problem solving (Consoli et al. 2016; Kline and Rosenberg 1986; Thomke, von Hippel, and Franke 1998; Vincenti 1990). Central to this perspective is the idea that innovations based on new scientific and technical knowledge derive from an inherently uncertain process of problem-solving, i.e. a series of repeated trials of potential solutions to a given problem. Such repeated trials lead to results which are used, in combination with insights about where possible solutions are to be found, in order to revise or refine the solutions, and the process continues until an acceptable result is reached. Thus, even if the process is uncertain the generation of variety is not blind as conceptualised in Darwinian evolution but rather Lamarckian as it is guided by insights by people as to where possible solutions are to be found (Hodgson 2015; McKelvey 1996).

But where do the insights come from that guide the actors involved in these repeated trials for possible solutions and generate variety? A first aspect to consider is the relationship to theory and practice. Following Fleming and Sorenson (2004) we propose that there are two major classes of insights that guide the search for possible solutions. First, there are *theory-driven* insights based on generalisable scientific knowledge, which provides the equivalence of a map that predicts the feasibility and efficacy of particular set of solutions. With the help of the map, the actors can engage in search for possible solutions, which can be directed towards the most promising parts of the map. Second, there are *experience-driven* insights based on previous experience. This includes experience from solving similar problems where the search proceeds in incremental steps and the feasibility and efficacy of the direction taken are difficult to predict in advance. A second aspect in the literature is that theory and practice may involve different epistemological communities. In this view, there is a distance between research as science, and practice as technology. The former focuses on the creation and validation of generalisable knowledge and the latter focuses more upon solving specific practical problems (Nightingale 1998, 2004). Thus, the implications are that bridging research and practice requires the combined insights from theory-driven search (informed by science) and experience-driven search (informed by practice).

In order to combine theory-driven and experience-driven problem-solving processes in our conceptual framework, we rely on one specific contribution which conceptualises problem-solving as engineering design (Vincenti 1990). Engineering design may be informed by science (theory-driven), but is also heavily dependent on practice (experience-driven). Furthermore, Vincenti (1990) argues that in engineering design, different means are used for testing and selecting the variety that is generated. On one hand, direct trials in real-world settings are used, e.g. flying in a full-sized prototype airplane. On the other hand, indirect trials are used where the complexity of real-world settings is reduced, e.g. by using a simplified, or a virtual, version of the test object (miniature airplane in a wind tunnel) or by decomposing the original problem into more controllable sub-problems (testing full-size wings in a wind tunnel). Simplified or virtual solutions provide an incomplete and indirect way of interacting with reality but reduce the number of possible solutions that need to be tried under real-world settings, which are less controlled and where failure tends to have more dire consequences (Campbell 1987; Thomke, von Hippel, and Franke 1998; Vincenti 1990).

Thus, for this second component, we propose that the generation and use of new scientific and technical knowledge – which is the subject of innovation governance – is conceptualised as an evolutionary problem-solving process operating at the intersection between research and practice. We mean that at this intersection – where there is true uncertainty about future outcomes – the generation of variety is guided by a combination of theory-driven and experience-driven insights with the selection of variety through both direct or indirect trials. Our interpretation is that because direct trials tend to be less controlled, and take place in existing systems of practice, they usually pose larger risks of negative unintended consequences as compared to indirect trials.

Our third component relates the two components above to the empirical setting of medical innovation. Here, we can follow extant literature on medical innovation, which provides an evolutionary perspective on the generation and use of scientific and technical knowledge. Specifically, this literature stresses how and why medical research and clinical practices operate as separate, but co-evolving, epistemic communities that, taken together, shape the growth of knowledge of medical technologies and their use in clinical practice (Consoli and Mina 2009; Gelijns and Rosenberg 1994; Metcalfe, James, and Mina 2005; Morlacchi and Nelson 2011; Rosenberg 2009). Thus, equivalent to the distance between science and technology in the general innovation studies literature, the medical innovation literature has identified the epistemological distance between medical research and clinical practice. Moreover, the above distinction we made between direct and indirect trials is especially salient for medical innovation because direct trials, such as clinical trials, involve humans while indirect trials do not.

Specifically, this medical innovation literature has what we consider three main approaches to studying how the epistemological distance between medical research and clinical practice may be bridged. One stream of literature stresses that bridging occurs at the organisation level, specifically hospitals. This literature emphasises the centrality of hospitals in medical innovation – generally with a positive effect on innovation – where innovations include a wide range of new medical devices, pharmaceuticals, and clinical procedures (Hopkins 2006; Lander and Atkinson-Grosjean 2011; Thune and Mina 2016). Another stream of literature stresses that bridging occurs at the individual level of the scientist-clinician. This literature focuses upon the role of clinician-scientists, that are jointly employed by hospitals and universities, and argue that these individuals help solves a perceived paradox of modern biomedicine, namely the limited impact that the molecular revolution has had on clinical practice (Lenfant 2003; Vignola-Gagné, Biegelbauer, and Lehner 2014). Finally, a third category of literature stresses the importance of clinical research as a set of activities. This literature is concerned that the prominence of theory-driven research in molecular biology and molecular genetics has led to a reduced emphasis on clinical research, which is more experience-driven and better connected to medical practice (Ahrens 1992; Gittelman 2016; Hirsch 1997). We combine the three approaches.

Thus, for this third component, we stress the importance of clinical research. We conceptualise that medical research and clinical practices operate as separate, but co-evolving, epistemic communities involved in the generation and use of new scientific and technical knowledge for medical innovation. These communities are bridged by clinician-scientists conducting clinical research in hospitals. Our rationale follows. The important role that hospitals play within science, technology and innovation processes is directly related to their organisational role as the venue for clinical research involving patients (direct trials). Such clinical research is performed by clinician-scientists, who may complement their research on patients in hospitals with laboratory research outside the hospital that involves computer simulations, animal models and cell systems instead of patients (indirect trials). Thus, we specify that individual clinician-scientists, through their joint appointment at a university and a hospital, are part of a collective action including both organisations. Through clinical research activities, the individual and organisational level meet in an attempt to generate and use new knowledge by bridging the epistemological communities of medical research and clinical practice in the context of the hospital and involving patients.

Taken together, we combine these three components into our conceptual framework, visualised in Figure 1.

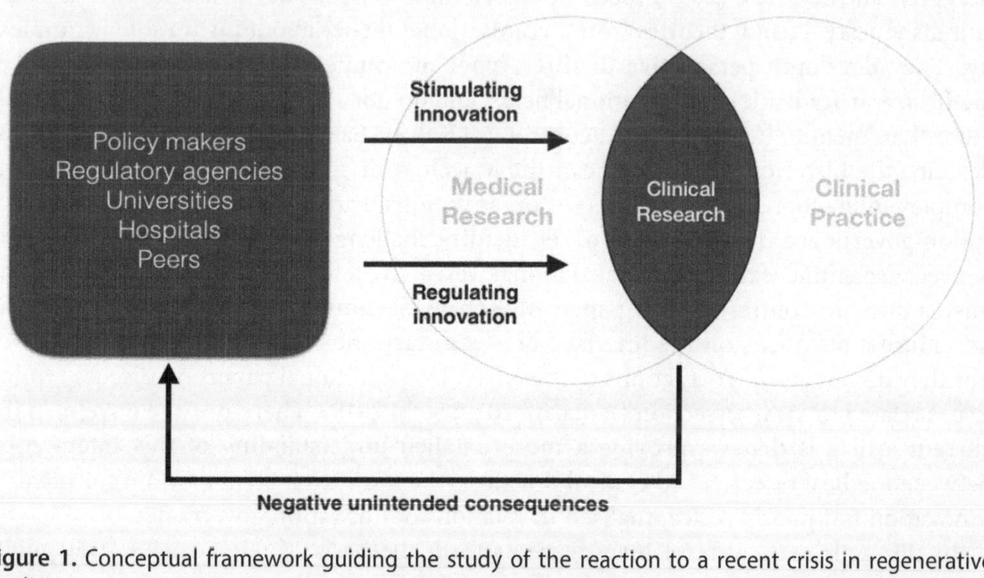

Figure 1. Conceptual framework guiding the study of the reaction to a recent crisis in regenerative medicine.

As shown in Figure 1, this paper is concerned with innovation governance for clinical research. We have defined clinical research narrowly, as the generation and use of new knowledge by clinician-scientists in the context of the hospital and involving patients. We conceptualise clinical research as related to both medical research and clinical practice, as shown by the overlap of the two circles. Therefore, the process of clinical research is seen as an evolutionary problem-solving process, where the generation of variety can be guided by theory-driven or experience-driven insights, but where selection is done through direct trials. This is regulated by the wider innovation governance system, which consists of policy makers, regulatory agencies, universities, hospitals and medical research peers. Innovation governance can include both the stimulation and regulation of clinical research activities, as represented by the two arrows in the middle. Furthermore, we propose that negative unintended consequences are not fully dealt with by existing regulation, which means that the system needs to react to them when they occur as represented by the arrow at bottom of the Figure.

3. Methodology

Our research design is a longitudinal single case study of a crisis related to innovation governance. We chose a single case study design because the selected case is both extreme and complex (Flyvbjerg 2006; Yin 1994). The case study chosen is extreme with regards to the magnitude of the reaction by the innovation governance system and provides a unique opportunity (as compared to other country and time-points) to analyse what each relevant actor specifies they consider as normal and deviant behaviour, what is contested, as well as what regulatory changes they propose. Furthermore, the case involves a complex system of governance and provides unusually rich information, which we analyse in relation to our objectives.

At the time of writing, there are two existing publications in our field which provide detailed, but different, information about other aspects of the Macchiarini crisis. Berggren and Karabag (2019) focus upon scientific misconduct – also known as fraud and dishonesty – using theories from organisational theory about institutional complexity. They develop a perspective of three types of competing logics (market-oriented, medical, and academic) of institutional fields, and do not address innovation governance, other than mentioning fragmented control. McKelvey, Saemundsson, and Zaring (2018) describe the historical development of the Macchiarini crisis, and explicitly analyse the complex interactions between different organisations using the theoretical lens of innovation governance. They do so in order identify challenges for public policy, balancing between scientific excellence, translational research, and opportunities for scientific misconduct. In contrast to this paper, McKelvey, Saemundsson, and Zaring (2018) do not address our focus on the interface between scientific research and clinical practice nor they do not consider reactions to the crisis. Two additional papers do discuss limited issues related to the intersection (Arnason 2019; Sethi 2019). Therefore, the novelty in the current article is that we provide a more detailed understanding of this intersection between medical research and clinical practice, which is extremely important for medical innovation but poorly conceptualised in relation to innovation governance.

In this article, the official investigations form the basis of our analysis. Ten public investigations were carried out, leading to 10 lengthy reports that represent the perspective of each actor in the innovation governance system. Under the Swedish public information law, all official investigations and material at authorities must be made available upon request, and many are posted on websites. While a few things are in English, the documents are primarily in Swedish, a language that both authors master and so translations below are done by the authors. We explicitly chose not to conduct interviews, partly due to the contested nature of the case, and partly due to the extensive nature of the written documentation.

We have gathered data during four years through an iterative process, with two separate steps of data gathering, also in relation to theoretical development.

The first step of the data collection began in January 2016, when both authors watched the TV documentary, which sparked a wider public interest in the scandal emerging around the clinician-scientist Pablo Macchiarini and the Karolinska Institute (KI). The authors immediately agreed to develop a joint IT-based retrieval system to systematically gather all documentation from 2010 to end of 2016 from: the university (Karolinska Institute, KI); the university hospital (Karolinska University Hospital, KuH), main research financiers (Swedish Research Council, Swedish Heart and Lung Fund); Retraction Watch; scientific journal articles mentioned in relation to accusations of scientific misconduct; influential national newspapers and magazines; as well as downloaded copies of websites from this group (due to the possibility they would be shut down). Based on this material, the authors jointly wrote up a thorough empirical description of the case, with approximately 100 pages of detailed description of the main events ordered chronologically. Based upon this chronological case, McKelvey, Saemundsson, and Zaring (2018) was published, constituting a detailed descriptive case on much more limited material than the current paper, as well as a different goal. Later, up until January 2019, we continued to gather press releases, articles in professional magazines and the popular press, and blog entries, which provided information about the 10 official investigations.

The second step of data collection for this paper specifically was to gather all official investigations (10) and related statements, which we have categorised as reactions to the crisis. As specified in the Appendix, the material from the investigations, reports and statements, consists of more than 1,000 pages of text. This documentation represents detailed information, and an unusually rich opportunity to map the perspectives. Each of these documents has slightly different foci, but all address the interface of scientific research and clinical practice. One author read all documentation, hand-coded, and categorised their statements as to 1) what activities they considered to constitute clinical research, i.e. generation and use of new scientific and technical knowledge at the intersection between medical research and clinical practice, 2) their investigation as to what went wrong, e.g. the negative unintended consequences, and, 3) their proposed regulatory improvements to avoid similar crises in the future. Both authors discussed the results and appropriate categories of analysis.

Through this critical reading of the sources, we reconstructed the sequence of events – and the involvement of each of the actors. We thereafter derived three phases relative to theory, which we call the stimulation of innovation, the generation of negative unintended consequences, and reaction by innovation governance actors, which we used to structure the presentation of the case study. Moreover, as an outcome, we categorise the activities constituting clinical research into three conceptual categories as search for new knowledge, search for new solutions, and support for new solutions. Finally, one author wrote up the initial analysis, and then new weekly versions were then jointly discussed by both authors, through weekly meetings over 18 months, and both authors engaged in revisions of the manuscript.

There are many limitations to our research design of a single case study, chosen as extreme for theoretical reasons. We do not claim that the case is representative nor valid across countries, nor even within Sweden over time. However, we have chosen it as an extreme case for theoretical sampling, as it provides new insights into the dynamics of innovation governance at the interface between medical research and clinical practice.

Although not the focus here, we acknowledge that there is a wider interest in the topic of fraud and inappropriate scientific conduct across fields of research (Hall and Martin 2019). The literature describes many similar crises related to scientific misconduct within regenerative medicine specifically (Cyranoski 2012; Kim and Park 2013; Bik, Cadadevall, and Fang 2016; Adam. 2019). Moreover, we acknowledge the expanding stream of research which addresses scientific misconduct – such as fraud, falsification, dishonesty, retraction, inappropriate image duplication and other dubious academic practices.

4. Results

We present and analyse our case study in two parts below. First, using categories we found in our case study, we briefly describe the crisis and summarise it as three phases: stimulation, unintended consequences, and reaction. Second, we specifically analyse the reaction of the innovation governance actors to the negative unintended consequences of the crisis, by focusing on the regulation of the interface between medical research and clinical practice.

4.1. The three phases of the crisis

Figure 2 gives an overview of the sequence of events (2008–2016) leading up to the crisis and the actions of the actors of involved with the governance of medical innovation, the

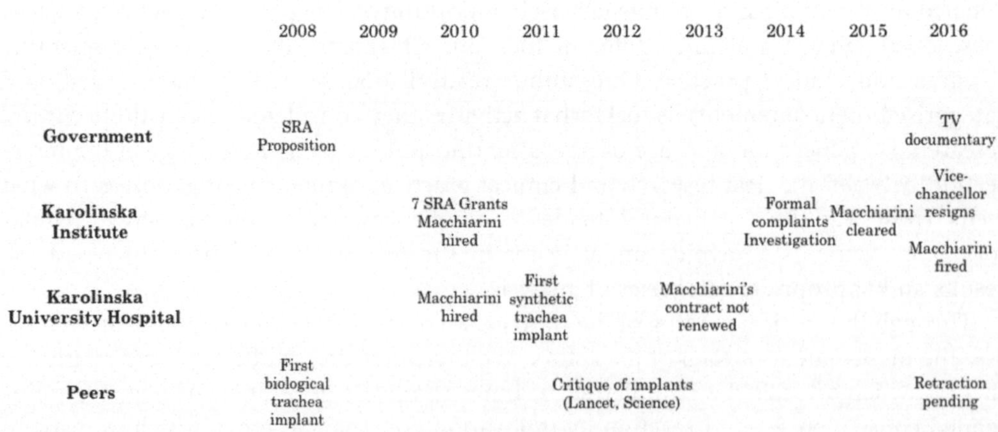

Figure 2. Overview of the sequence of events leading to the Macchiarini crisis and actions of actors involved in the governance of medical innovation.

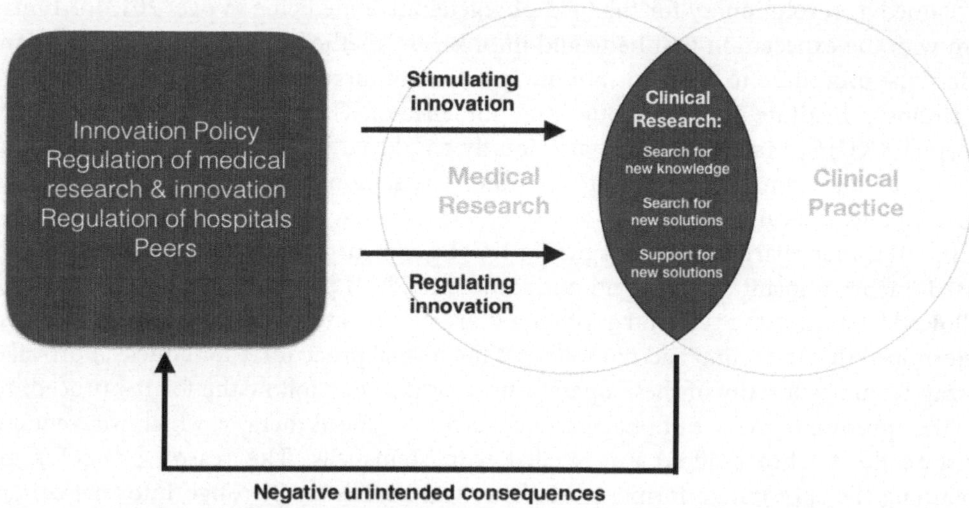

Figure 3. Specification of clinical research activities.

Government, the Karolinska Institute (KI), the Karolinska University Hospital (KUH), and research peers.

The origin of the crisis can be found in a very large public policy initiative to promote excellent science. When presenting their annual research bill in 2008, the Swedish government initiated a new funding for selected Strategic Research Areas (SRA). The objective of this initiative was to increase international competitiveness of Swedish industries as well as to produce the highest international excellence in science (Swedish Executive Government 2008). For the total SRA policy initiative, the government invested a total of 590 million EUR between 2010 and 2014 (Swedish Research Council 2015).

In 2010–2014 the Karolinska Institute (KI) received money from the SRA initiative – one of seven SRA grants obtained by KI – for a 'strategic research program in Stem Cell Research and Regenerative Medicine' that 'supports research that advances our understanding of stem cell biology and approaches to bring regenerative medicine to the clinic, for future treatment of diseases for which there currently are no therapies' (StratRegen 2016). We estimate that during the period 2010–2014, the KI research programme received a total of around 15,5 million EUR. Furthermore, KI used the SRA program as a platform to obtain a number of additional large grants in regenerative medicine from various other sources – mostly public ones but also foundations – in order to establish and support a number of research centres in the area. In total, we estimate that KI obtained and spent between 33 and 50 million EUR on regenerative medicine between 2010 and 2014.

As a way to realise the KI ambition to achieve the highest international excellence in regenerative medicine, KI recruited the clinician-scientist Paolo Macchiarini in 2010 and also established the Advanced Centre of Translational Regenerative Medicine (ACTERM) research centre for him to lead. One reason for recruiting him was that Macchiarini and his colleagues had carried out a human trachea transplant operation in 2008 using stem cells and a biological implant. Their operation was novel, and widely acclaimed as revolutionary for the field of regenerative medicine (Vogel 2013). KI hired him with the expectation that he would improve his method for airway transplants and 'adapt the procedure to other intrathoracic organs of increasingly complex architecture' (Karolinska Institute 2016). Through the influence of KI on the Karolinska University Hospital (KUH), Macchiarini was also jointly employed by the hospital. KUH emphasised that his primary role was to conduct translational research bridging medical research and clinical practice.

In 2011, Macchiarini and colleagues at KUH performed the first of its kind stem cell-based trachea implant using a synthetic scaffold made of polymers. One more operation followed later the same year and another a year later. In later public investigations, it was determined that the group had not followed the formal procedures for ethical approval of research studies for any of these operations. Nor did they follow the formal procedures for the approval of the use of pharmaceuticals for advanced therapy, which was required because the trachea scaffold was seeded with stem cells. The reason given for not obtaining the appropriate formal procedures was as follows. Through informal contact with regulatory agencies and internal conferences at KUH, the clinical managers made the judgement that the decision to operate could be based on the patient's critical condition and on the lack of alternatives, and therefore did not require formal regulatory approval (Asplund 2016). However, the operation was used for medical research. Despite not obtaining permission for a research study, the authors later wrote up the operation of the first patient as the subject of a research publication (Jungebluth et al. 2011). Moreover, KI reported the results to the government, and specified that it considered these operations to be highly successful and also an outcome of the KI regenerative research program (Swedish Research Council 2012). Later investigations showed that the involved physicians did not follow professional guidelines, such as the Helsinki declaration on ethical principles for medical research involving human subjects. This declaration states that even if unproven interventions may be used in the hope of 'saving life, re-establishing health or alleviating suffering' they should not be repeated, or if done subsequently, where possible, be made the object of research (WMA 2008). Therefore, according to these guidelines, at least the two latter operations should have been a part of an approved research project.

In parallel, scientific misconduct was quickly alleged. The operations for the trachea implants were initially hailed and reported in journal articles as very successful. However, some research peers argued that it was impossible that the implants were working as well as reported in the 2011 paper. Several things happened in relation to these allegations. Macchiarini's contract with KUH was not renewed in 2013, despite pressures from KI to continue the joint appointment (Asplund 2016). Moreover, in 2014 four physicians jointly employed by KI and KUH filed a formal complaint about scientific misconduct at KI, suggesting that scientific papers authored by Macchiarini had incorrectly described the benefits of the implants. In 2014 a Belgian professor also filed a formal complaint to KI accusing Macchiarini for scientific misconduct. Following an internal and external investigation of the allegations – as required by university regulation – KI freed Macchiarini of scientific misconduct in 2015 (Hamsten and Samuelsson 2015a, 2015b). At this point, the internal university decision to clear his name was taken, despite the fact that the external examiner (Gerdin 2015) had been very critical in his report.

In early 2016 the Swedish public television (SVT) aired a three-hour documentary called 'The Experiments' (Experimenten), filmed by Bo Lindquist (2016). The documentary followed Macchiarini over several years as he implanted synthetic tracheas at KUH and in Russia, and presents information that suggested that the operations lacked proper scientific support and regulatory approval. Furthermore, it showed leaders of the prestigious KI defending the clinician-scientist's conduct despite mounting evidence of its inappropriateness. A public outcry followed in Swedish media. After initially supporting him, and following various resignations from prestigious posts, KI finally decided to relieve Macchiarini of his duties and close down the ACTREM centre. These interlinked events initiated a series of 10 public investigations (see Appendix for a complete list) to restore confidence in the governance of medical research and innovation in clinical practice.

Based on our analysis we summarise the crisis as three phases of stimulation, negative unintended consequences, and reaction, as seen in Table 1.

Table 1 shows the details for each actor in the innovation governance system. The first phase is characterised by the stimulus from the major funding of regenerative medicine in Sweden, which is initiated and led by the government, and high expectation of the clinical value of regenerative medicine in general and Macchiarini's research in particular. The second phase is characterised by the emergence of the negative unintended consequences of the stimulus. These appear as the execution of an unsuccessful high-risk clinical procedure that was made possible by scientific misconduct and the bypassing of regulatory procedures and ethical guidelines, and further amplified by inappropriate handling of allegations of misconduct. Finally, the third phase is characterised by the reactions of the actors involved in the governance of medical innovation, once the TV documentary had resulted in a public outcry. In the next section, we analyse these reactions in more detail with a focus on the regulation of the interface between medical research and clinical practice.

4.2. Reacting to negative unintended consequences

To regain public confidence in the governance of medical innovation, official investigations were initiated that focused on what had happened, why it had happened, and how it could be avoided in the future. Our focus is on the intersection between medical research and clinical practice, i.e. clinical research, which we have defined as the generation and

Table 1. The Macchiarini crisis summarised as three phases of stimulation, unintended negative consequences, and reactions and how each phase relates to the actors in the innovation governance.

Actor	Stimulation	Unintended negative consequences	Reaction
Government	SRA funding.	Bypassing of regulations.	Investigation (2,8,9,10).
Karolinska Institute (KI)	Research centre.	Inappropriate recruitment process Influence on KUH. Scientific misconduct. Handling of allegations.	Investigation (1,4,6,7)
Karolinska University Hospital (KUH)	KI influence.	New high-risk clinical practice.	Investigation (3)
Peers	First biological trachea implant.	Bypassing of ethical guidelines.	Investigation (5)

use of new knowledge by clinical-scientists in the context of the hospital and involving patients. Thus, we focus on how – in the investigations – the activities constituting clinical research were defined, how they were supposed to be regulated according to existing regulation, and what changes were suggested to avoid similar crises in the future.

Most of the investigation reports refer to the same sources when defining what activities constitute clinical research. On one hand, they refer to the Health and Medical Services Act (Hälso- and sjukvårdslagen) and the Patient Safety Act (Patientsäkerhetslagen) for defining clinical practice and its relationship to science. On the other hand, they refer to the Ethical Review Act (Etikprövningslagen) when defining research. Furthermore, they relate these concepts to other concepts used by the medical profession. Some of these concepts are concerned with activities performed by clinicians and scientists, but others are concerned with exceptions, i.e. situations under which standard regulations do not apply (Table 2).

According to Swedish law and regulation, clinical practice is defined as medical measures – based on science and confirmed experience ('vetenskap och beprövad erfarenhet') – that are used to prevent, diagnose and treat diseases and injuries (Asplund 2016; Gerdin 2015; Heckscher, Carlberg, and Gahmberg 2016; Lindvall and Engström 2016; SMER 2016; SOU 2017). The concept of confirmed experience is not defined in the law nor is it an internationally established concept. Thus, many of the reports discuss the boundaries of confirmed experience in order to identify the boundaries between clinical practice and medical research.

Lindvall and Engström (2016), who wrote a report on behalf of The Swedish Society of Medicine and the Royal Academy of Sciences, use the concept of non-confirmed treatment ('obeprövade behandlingsmetoder') and medical innovation for measures that physicians historically have used to treat seriously ill patients, which, despite being often based on science, cannot be seen as confirmed by experience. These measures may be early in their development or may have been used for other patient groups or indications. The rationale for using these measures in clinical practice has been the seriousness of the illness and the lack of alternative treatment options. Similarly, SMER (2016) defines innovative therapy as a clinical procedure that is being used without its benefits and risks being evaluated in clinical trials and whose efficacy has not been confirmed by experience. Innovative therapy, or non-confirmed treatment, are mentioned in most of the reports, usually with a reference to the Helsinki declaration and guidelines by the International Society for Stem Cell Research for the existence and justification of such procedures.

Table 2. Concepts used by the medical profession and Swedish law and regulations, as reported by investigation reports in the wake of the Macchiarini crisis, to define activities at the intersection between medical research and clinical practice.

Not defined by Swedish law and regulation	Vital indication. Compassionate use.	Non-confirmed treatment. Medical innovation. Innovative therapy. Clinical research. Translational research.
Defined by Swedish law and regulation	Hospital exception. Compassionate use program.	Clinical practice. Clinical trials. Research.
	Exceptions	Activities

The exceptions are usually reviewed in the investigation reports as possible justification for the use of non-confirmed treatment or innovative therapy. The hospital exception and the compassionate use program are specifically concerned with the possibility to produce and use pharmaceuticals that are still in development and both require approval from the Medical Products Agency. Vital indication and compassionate use are ethical principles justifying the use of unusual measures in urgent situations (vital indication) or when no other known alternatives exist (compassionate use). However, according to Asplund (2016), these principles cannot be used in order to bypass regulation.

The concept of clinical trials is defined in Swedish law and regulation as concerned with a clinical investigation of the effects of the use of a pharmaceutical or a medical device on humans or animals (SOU 2017). Lindvall and Engström (2016) define clinical trials specifically in the context of pharmaceuticals as the study of the efficacy and safety of a pharmaceutical. For pharmaceuticals clinical trials are always required and, in some cases, they are required for medical devices. Knowledge about the benefits and risks of the pharmaceutical, or the device, is increased during the process, which is required before they are used in clinical practice (Asplund 2016).

According to Swedish law and regulation, medical research is defined as two types of activities. First, as systematic experimental or theoretical activities that have the aim of generating new knowledge. Second, as science-based development activities. Lindvall and Engström (2016) further define clinical research as research that involves patients, animals, or cell systems with the aim to generate scientific results useful for the development of new diagnostic methods or therapies that can solve a health problem, or identify factors that can improve health. Asplund (2016) does not provide an explicit definition of clinical research, but argues that when clinical research involves patients it becomes a mix of clinical practice and research. Both Lindvall and Engström (2016) and Asplund (2016) stress that the aim of clinical research involving patients is to generate new knowledge that can be applied to groups of patients as opposed to individual patients. Finally, Heckscher, Carlberg, and Gahmberg (2016) refer to translational research as research that aims to improve knowledge flows between research and practice in order to increase patient benefits.

After analysing how activities related to clinical research are defined in different investigate reports we make the following observations. First, there is consensus that clinical practice based on confirmed experience is not considered a part of clinical research. Also, there is consensus that research involving human subjects at the hospital is clinical research. Second, the goals of what research should lead to are expressed in various ways. In some cases, the generation of new knowledge is expressed as a goal in itself, whereas in other cases the goal is expressed as to increase patient benefits or to develop new or improved treatments. Moreover, in some cases the goal is expressed as to provide scientific support for the efficacy of a certain treatment, e.g. use of a particular pharmaceutical. Third, we interpret that there is ambiguity around the regulation of activities that are variously referred to as the use of non-confirmed treatment, medical innovation or innovative therapies. In Table 2, they are in the top, right-hand quadrant. These are activities that professionals believe have played an important role for progress in medicine, and should be within the boundaries of clinical research, but are not explicitly acknowledged by Swedish law and regulation.

Turning to the regulatory changes suggested by the authors of the investigation reports to restore confidence and avoid similar crises in the future, we focus on those directly related to clinical research in hospitals.[1] Our interpretation is that the majority of the suggested changes are related to the use of innovative therapies in extreme cases with the aim to specify more clearly than before under which conditions innovative therapies may be used and what decision-making procedures are required. While the decision to consider the use of an innovative therapy as a treatment option is supposed to originate from physicians, and be supported by patients' informed consent, the use of the therapy must also be supported by scientific knowledge and other medical professionals. Furthermore, the scientific rationale, along with evaluation of benefits and risks, needs to be documented and approved, first by clinical management and then by an external regulatory entity that specialises in reviewing applications for the use of innovative therapies. An exception is allowed if the health of the patient is likely to quickly deteriorate, in which case an application for review should be sent afterwards. Once an innovative therapy has been used for the first time, its further use is contingent upon the creation of a research study, subject to regulatory requirements of such a study.

The focus of the regulatory changes reported above on the use of innovative therapies reflects what the investigators identified as a major issue related to clinical research in hospitals, namely the justification for direct trials on human subjects. For each of the three patients, Macchiarini and his colleagues justified their decision to perform the surgery based on the ethics of clinical practice rather than the ethics of research. The investigators did not agree and their conclusion was based on the fact that Swedish law and regulation did not allow for the use of non-confirmed treatments as a part of clinical practice even in extreme cases, thus requiring ethical approval according to regulation concerning research on human subjects. The investigators did not propose any major change to the regulation of research on human subjects, because they – implicitly rather than explicitly – seemed to assume that the unintended negative consequences could have been avoided if the regulation was correctly applied. The regulatory process would have discovered the weak scientific basis for the operations and not approved the operations until the scientific evidence was strong enough.[2] However, most of the investigators, e.g. Lindvall and Engström (2016), SMER (2016) and SOU (2017), acknowledge that the use of non-confirmed treatments in extreme cases is a legitimate activity at the interface between medical research and clinical practice and should be accounted for in Swedish law and regulation. However, it should be subjected to external regulatory approval and not be applied more than once unless as a part of a research study.

5. Revisiting the conceptual framework

Having presented the case study of the crisis as three phases of stimulation, unintended consequences, and reaction and also analysed the reaction, we now return to the conceptual framework presented in Section 2.

[1]Other suggestions for regulatory changes that are not directly related to clinical research activities in hospitals concern routines for recruitment of clinician-scientists, processes for handling allegations of scientific misconduct, documentation of research activities, delegation of authority, and clarity in responsibility of joint activities by universities and hospitals.

[2]This assumption can be questioned in hindsight as the authors of scientific publications of results from animal studies preceeding the operations have been found guilty of scientific misconduct by providing misleading presentation, interpretation and description of the results and failing to present raw data on which these results were based (see e.g. CERB (2016)).

In the framework, we defined clinical research as the generation and use of new knowledge by clinician-scientists in the context of the hospital and involving patients. Furthermore, we conceptualised clinical research as an evolutionary problem-solving process, where the generation of variety is guided by theory-driven or experience-driven search, but where selection is done through direct trials, i.e. real-world settings involving patients. We defined innovation governance to include both the stimulation and regulation of clinical research activities. Furthermore, we expected that negative unintended consequences were not fully dealt with by existing regulation due to the tentative nature of innovation governance in the context of emerging science and technology.

Presenting the crisis as three phases has been useful for understanding the tentative nature of innovation governance and the challenges associated with the balancing of stimulation and regulation. Regenerative medicine is 'a field of medicine devoted to treatments in which stem cells are induced to differentiate into the specific cell type required to repair damaged or destroyed cell populations or tissues' (NIH 2015:23). One of the applications of regenerative medicine is the replacement of failed organs by artificial organs – as done by Macchiarini and his colleagues at the Karolinska University Hospital. In the last century a large body of scientific and technical knowledge was built about human organ transplants and such transplants are today performed as routine practice in many hospitals around the world. Regenerative medicine is an alternative to human transplants, but is yet to become a routine clinical practice. Instead, regenerative medicine is a fast-moving international field of research and innovation fuelled by large-scale funding (Coccia 2014; Salter and Faulkner 2011; Salter and Salter 2010). Thus, regenerative is a field where governments around the world – including the Swedish government – are stimulating research and innovation, while at the same time there are ambiguities in terms of how research and innovation should be regulated. We suggest that these ambiguities, in turn, generate negative unintended consequences and a reaction by the innovation governance system to restore confidence and avoid similar situations in the future. In this case study, the reaction is to specify more explicitly how clinical research is to be regulated and the role of external regulators. Their responses can be interpreted in two ways. It can be interpreted as stricter regulation because it extends the role of external regulators, but it can also be interpreted as a way to legitimise activities that are considered important means for technological innovation.

We find that the ambiguities in the regulation of clinical research are not evenly distributed among the activities that constitute clinical research. To elucidate these differences, we propose that clinical research is divided into three types of activities (Figure 3):

(1) *Search for new knowledge* aims to increase our basic understanding of life and diseases without necessarily searching for new treatment options. This activity is primarily theory-driven, i.e. informed by science, and the innovation governance actors agreed it should be regulated as research activity.
(2) *Search for new solutions* aims to increase patients' benefits or develop new or improved treatments. This activity is primarily experience-driven, i.e. informed by practice, but many of the innovation governance actors argued that the regulation

of these activities is not clear, e.g. how decisions are made about the use of innovative therapies in extreme cases.

(3) *Support for new solutions* aims to test and compare the efficacy of particular pharmaceuticals, medical devices, or treatment procedures. This activity is primarily theory-driven and the innovation governance actors agreed that this activity is either regulated as research or by specific regulation concerning clinical trials.

Our proposed division of clinical research into search for new knowledge, search for new solutions, and support for new solutions, highlights the difference between experience-driven and theory-driven search activities. While there is a high degree of consensus about the regulation of search for new knowledge and support for new solutions, which are primarily theory-driven activities, we found less consensus about the regulation of the search for new solutions, which is primarily experience-driven. Most innovation governance actors agree that the regulation does not clearly specify if, and under what conditions search for new solutions may be applied, even if clinician-scientists seem to agree about the importance of such activities for (past) progress in medicine (Ahrens 1992; Asplund 2016; Hirsch 1997). This ambiguity seems to have normalised action within the medical profession, either intentionally or unintentionally, that appeared to outsiders as deviant (Hedgecoe 2013). This likely creates the unintended consequences that generated the public outcry. In response, the regulation of clinical research was adjusted, in this case study, by more clearly specifying under what conditions search for new solutions may be used and how the decisions to use them are reviewed by actors in the innovation governance system.

6. Conclusions and future research

The purpose of this article is to better understand the challenges involved in avoiding the dark side of technological innovation processes. Using a longitudinal case study of medical innovation, we have explored how actors in the innovation governance system counterbalance the stimulation of novelty for future benefits for society with the regulation of novelty, in order to avoid the risk of unintended consequences. We analysed how actors in the respective innovation governance system reacted to negative unintended consequences; a reaction that we interpret as an attempt to restore the balance between the stimulation and regulation of technological innovation processes by clarifying ambiguities in the regulation at interface between research and practice.

Our first set of conclusions is concerned with the impacts of the ambiguities in the regulation of the interface between research and practice. We propose to conceptualise these ambiguities as *grey zones*. We define grey zones as situations when it is unclear if the benefits of experimentation – direct trials – outweigh its risks. We argue that grey zones are connected to the activities at the interface between research and practice, and specifically those we conceptualise as search for new solutions. Search for new solutions is characterised by experimentation in real-world settings which may include humans as well as socio-technical systems whose continuous operation is important for human well-being. The mere possibility of serious harm creates ambiguity about the level of risk analysis, and the scope of the expected benefits, that are needed in advance to justify the search. This ambiguity is further amplified by the different perspectives of research and practice when it comes to the beneficiaries and risk takers in the context of a new practice. For practice, the focus is on the benefits and risks of single constituents, while for research the benefits tend to be more general for the same level of risk. For example, from the perspective of clinical practice, the unique situation of each patient – due to anatomical variations, condition of the disease, or available methods and resources for treatment – needs to be considered when selecting treatment options. If the benefits outweigh the risks for the individual patient, then deviations from standard practice, such as doses of approved pharmaceuticals, use of approved pharmaceutical for different indications than originally intended, and new uses of approved medical devices, are accepted (Schwartz 2014). From the perspective of medical research – where the objective is to search for generalised knowledge – the benefits are much larger for the same level of risk as the beneficiaries also include all future patients that would benefit from the new practice.

Moreover, we propose that grey zones are continually created and resolved over time, as innovation governance systems counterbalance innovation and the risk of negative unintended consequences. We argue that the main reason for the continual existence of grey zones is the inherent uncertainty of technological innovation, as generated through an inherently uncertain process of problem solving (Consoli et al. 2016; Kline and Rosenberg 1986; Thomke, von Hippel, and Franke 1998; Vincenti 1990). When a particular class of problems emerges (e.g. cancer) different classes of solutions are developed to solve them (e.g. surgery, radiation therapy, and chemical therapy). Search at the interface between research and practice generates new scientific and technological knowledge related to the nature of the problem and the efficacy of the different sets of solutions. This includes knowledge about the benefits and risks of applying a particular class of solutions to a particular class of problems, which will reduce grey zones. However, new classes of problems will always emerge as well as new classes of solutions that can be applied to both new and existing classes of problems. For these new classes of solutions there may be great expectations about their potential benefits, but – at least to begin with – limited knowledge about how these benefits are realised. Furthermore, there will also be incomplete knowledge about the risks that need to be taken to develop the knowledge and artefacts for consistent and reliable use of the solutions in practice and the risks of repeated use. Thus, inherent uncertainty about future problems and solutions means that, even if some grey zones are resolved, new ones will emerge along with emerging science and technology.

Our second set of conclusions is for medical innovation specifically. There have been longstanding debates in medicine about the relative effectiveness of different means for generating and selecting new clinical practices. Already in the early 20th century, when the Rockefeller Institute of Medical Research and the associated Rockefeller Hospital were established, physicians debated whether ideas generated by scientists at the laboratories of the Institute should be tested by physicians at the hospital or if physicians working at the hospital should generate ideas through observing and measuring patients at the bedside followed by investigations in the laboratories before being confirmed by application in the hospital (Hirsch 1997). Recently, observers including Gittelman (2016), have associated the progress in 20th century medicine to the experience-driven, approach and the current slowdown of progress to the dominance of a theory-driven approach to innovation. We believe that our conceptual model and interpretations from our case study provide an opportunity to bring a more nuanced perspective to this debate.

In contrast to the arguments about a shift in beliefs about the relative effectiveness of the two approaches put forth by Gittelman (2016), we propose that the current prominence of theory-driven approach to medical research and innovation may be better explained by the role of the innovation governance system in shaping the means available for generating and applying new knowledge at the intersection between medical research and clinical practice. Even if experience-driven activities were more effective in generating and selecting among ideas of new clinical practices, their effectiveness is counterbalanced by their potential for negative unintended consequences. When unintended consequences emerge, they will lead to reactions by actors in the innovation governance system, which may subsequently make experience-based activities less available as means for generating and applying new knowledge due to more stringent external regulation. An interesting opportunity for further research is to test this proposition by empirically studying changes in innovation governance of medical innovation in selected fields as multiple phases of stimulation, negative unintended consequence, and reaction.

Our final set of conclusions relate to the more general debate about the intersection and relationships between scientific research and technological innovation, and, hence, the relationship between theory-driven search and experience-driven search. Nightingale (2004) argues that it is difficult, and often impossible, to accurately predict complex phenomena from first principles, even if they may be predicted through experience of empirical regularities. Thus, it is difficult, and often impossible, to develop new practice without experience-driven search and direct trials, even if the new practice is inspired and guided by scientific research. At the same time, the increasing power of predictive science (Arora and Gambardella 1994; Fleming and Sorenson 2004) suggests that the importance of theory-driven indirect trials for technological innovation is increasing reducing the need for experience-driven direct trials. We propose that the continual ambiguity around searches for new solutions at the interface between research and practice – primarily an experience-driven activity using direct trials – provides pressure for using indirect trials, such as computer simulation, because there will be less risk for negative unintended consequences. These pressures will be stronger in times of crisis and in those real-world settings where the risks of unintended consequences are higher.

Acknowledgments

This research was financed by the Swedish Research Council Distinguished Professor's Programme, awarded to Professor McKelvey, on "Knowledge-intensive Entrepreneurial Ecosystems: Transforming society through knowledge, innovation and entrepreneurship", VR DNR 2017-03360.

This research was also financed by the Bank of Sweden Tercentenary Foundation (Riksbankensjubileumsfond) through the project "How Engineering Science Can Impact Industry in a Global World", lead by Professor M. McKelvey (FSK15 1080 1). This project is part of a large research program "The Long Term Provision of Knowledge" financed jointly by the Bank of Sweden Tercentenary Foundation, Formas, Forte and the Swedish Research Council.

Disclosure statement

No potential conflict of interest was reported by the authors.

ORCID

Maureen McKelvey http://orcid.org/0000-0002-1457-7922
Rögnvaldur J. Saemundsson http://orcid.org/0000-0002-8563-9078

References

Adam., D. 2019. "How a Data Detective Exposes Suspicious Medical Trials." *Scientific American. August* 6: 2019.

Ahrens, E. 1992. *The Crisis in Clinical Research*. Oxford: Oxford University Press.

Arnason, G. 2019. "Regulating Clinical Innovation: Trachea Transplants and Tissue Engineering." *American Journal of Bioethics* 19 (6): 32–34.

Arora, A., and A. Gambardella. 1994. "The Changing Technology of Technological Change: General and Abstract Knowledge and the Division of Innovative Labour." *Research Policy* 48 (2): 428–443.

Asplund, K. 2016. *The Macchiarini Case*. Stockholm: Karolinska University Hospital (In Swedish.

Berggren, C., and S. F. Karabag. 2019. "Scientific Misconduct at an Elite Medical Institute: The Role of Competing Institutional Logics and Fragmented Control." *Research Policy* 23: 523–532.

Bik, E., A. Cadadevall, and F. Fang. 2016. "The Prevalence of Inappropriate Image Duplication in Biomedical Research Publications." *MBio* 7 (3): 1–8.

Borrás, S., and J. Edler, eds. 2014. *The Governance of Socio-Technical Systems. Explaining Change*. Cheltenham: Edward Elgar Publishing.

Borrás, S., and C. Edquist. 2019. *Holistic Innovation Policy: Theoretical Foundations, Policy Problems and Instrument Choices*. Oxford: Oxford University Press.

Campbell, D. 1987. "Evolutionary Epistemology." In *Evolutionary Epistemology, Rationality, and the Sociology of Knowledge*, edited by G. Radnitzky and W. W. Bartley, 47–89. Chicago: Open Court.

CERB. 2016. "Statement." Accessed 24 April 2018. https://www.onep.se/media/2374/o-1-2016-statement-expert-group-for-misconduct-in-research-160906-eng.pdf

Coccia, M. 2014. "Emerging Technological Trajectories of Tissue Engineering and the Critical Directions in Cartilage Regenerative Medicine." *International Journal of Healthcare Technology and Management* 14 (3): 194–208.

Consoli, D., and A. Mina. 2009. "An Evolutionary Perspective on Health Innovation Systems." *Journal of Evolutionary Economics* 19 (2): 297–319.

Consoli, D., A. Mina, R. Nelson, and R. Ramlogan, eds. 2016. *Medical Innovation. Science, Technology and Practice*. London: Routledge.

Cyranoski, D. 2012. "Stem-cell Fraud Hits Febrile Field: After Heart-treatment Claims Collapse, Researchers Caution against a Rush to the Clinic." *Nature:News* 16 (October): 2012.

Edler, J., and J. Fagerberg. 2017. "Innovation Policy: What, Why and How." *Oxford Review of Economic Policy* 33 (1): 2–23.

Fagerberg, J., D. C. Mowery, and R. R. Nelson, eds. 2005. *The Oxford Handbook of Innovation*. Oxford: Oxford University Press.

Fleming, L., and O. Sorenson. 2004. "Science as a Map in Technological Search." *Strategic Management Journal* 25: 909–928.

Flyvbjerg, B. 2006. "Five Misunderstandings About Case-Study Research." *Qualitative Inquiry* 12 (2): 219–245.

Gelijns, A., and N. Rosenberg. 1994. "The Dynamics of Technological Change in Medicine." *Health Affairs* 13: 28–46.

Gerdin, B. 2015. "Särskilt Yttrande I Ärende Dnr: 2-2184/2014." (in Swedish). Accessed 24 May 2016. http://ki.se/sites/default/files/externa_granskarens_slutliga_rapport.pdf

Gittelman, M. 2016. "The Revolution Re-visited: Clinical and Genetics Research Paradigms and the Productivity Paradox in Drug Discovery." *Research Policy* 45: 1570–1585.

Hall, J., and B. Martin. 2019. "Towards a Taxonomy of Research Misconduct: The Case of Business School Research." *Research Policy* 48 (2): 414–427.

Hamsten, A., and L. Samuelsson 2015a. "Regarding Accusations of Research Misconduct by Paole Macchiarini." Dnr: 2-2184/2014, Decision 1, August 28. Stockholm: Karolinska Institute (in Swedish).

Hamsten, A., and L. Samuelsson 2015b. "Regarding Accusations of Research Misconduct by Paole Macchiarini." Dnr: 2-2184/2014, Decision 2, August 28. Stockholm: Karolinska Institute (in Swedish).

Heckscher, S., I. Carlberg, and C. Gahmberg. 2016. *The Karolinska Institute and the Macchiarini Case. An External Investigation.* Stockholm: Karolinska Institute (in Swedish).

Hedgecoe, A. 2013. "A Deviation from Standard Design? Clinical Trials, Research Ethics Committees, and the Regulatory Co-construction of Organizational Deviance." *Social Studies of Science* 44/1: 59–81.

Hirsch, J. 1997. "The Role of Clinical Investigation in Medicine: Historical Perspective from the Rockefeller University." *Perspectives in Biology and Medicine* 41 (1): 108–117.

Hodgson, G. M. 2015. *Conceptualizing Capitalism: Institutions, Evolution, Future.* Chicago: University of Chicago Press.

Hopkins, M. 2006. "The Hidden Research System: The Evolution of Cytogenetic Testing in the National Health Service." *Science as Culture* 15 (3): 253–276.

Jungebluth, P., et al. 2011. "Tracheobronchial Transplantation with a Stem-cell-seeded Bioartificial Nanocomposite: A Proof-of-concept Study." *The Lancet* 378: 1997–2004.

Karolinska Institute. 2016. "Paolo Macchiarini, guest professor." Accessed 21 May 2016. http://130.237.126.252/?page_id=81

Kim, J., and K. Park. 2013. "Ethical Modernization: Research Misconduct and Research Ethics Reforms in Korea following the Hwang Affair." *Scientific and Engineering Ethics* 19 (2): 355–380.

Kline, S. J., and N. Rosenberg. 1986. "An Overview of Innovation." In *The Positive Sum Strategy: Harnessing Technology for Economic Growth*, edited by R. Landau and N. Rosenberg, 275–305. Washington, DC: National Academies Press.

Kuhlmann, S., P. Stegmaier, and K. Konrad. 2019. "The Tentative Governance of Emerging Science and technology—A Conceptual Introduction." *Research Policy* 48: 1091–1097.

Lander, B., and J. Atkinson-Grosjean. 2011. "Translational Science and the Hidden Research System in Universities and Academic Hospitals: A Case Study." *Social Science & Medicine* 72: 537–544.

Lenfant, C. 2003. "Clinical Research to Clinical Practice—lost in Transition?" *New England Journal of Medicine* 349 (9): 868–874.

Lindquist, B. 2016. *Experimenten [Television Documentary in Three Parts].* Stockholm: SVT (in Swedish. January.

Lindvall, O., and I. Engström. 2016. *Clinical Guidelines for the Use of Non-confirmed Therapy for Severely Ill Patients.* Stockholm: Royal Swedish Academy of Sciences and the Swedish Society of Medicine (in Swedish).

Mazzucato, M. 2018. "Mission-oriented Innovation Policies: Challenges and Opportunities." *Industrial and Corporate Change* 27 (5): 803–815.

McKelvey, M. 1996. *Evolutionary Innovations.* Oxford: Oxford University Press.

McKelvey, M., R. Saemundsson, and O. Zaring. 2018. "A Recent Crisis in Regenerative Medicine: Analyzing Governance in order to Identify Public Policy Issues." *Science and Public Policy* 45 (5): 608–620.

McKelvey, M., O. Zaring, and S. Szücs. 2019. "Conceptualizing Evolutionary Governance Routines: Governance at the Interface of Science and Technology with Knowledge-intensive Innovative Entrepreneurship." *Journal of Evolutionary Economics* Published online 5: 2019. January.

Metcalfe, S., A. James, and A. Mina. 2005. "Emergent Innovation Systems and the Delivery of Clinical Services: The Case of Intra-ocular Lenses." *Research Policy* 34 (9): 1283–1304.

Moreno, J. D. 2001. *Undue Risk. Secret State Experiments on Humans.* New York: Routledge.

Morlacchi, P., and R. R. Nelson. 2011. "How Medical Practice Evolves: Learning to Treat Failing Hearts with an Implantable Device." *Research Policy* 40 (4): 511–525.

Nelson, R. R., and S. G. Winter. 1982. *An Evolutionary Theory of Economic Change*. Cambridge, MA: Harvard University Press.

Nightingale, P. 1998. "A Cognitive Model of Innovation." *Research Policy* 27: 689–709.

Nightingale, P. 2004. "Technological Capabilities, Invisible Infrastructure and the Un-social Construction of Predictability: The Overlooked Fixed Costs of Useful Research." *Research Policy* 33: 1259–1284.

NIH. 2015 8 April. 'Stem Cell Basics'. Accessed 10 March 2016. http://stemcells.nih.gov/staticre sour ces/info/basics/SCprimer2009.pdf

Rosenberg, N. 2009. "Some Critical Episodes in the Progress of Medical Innovation: An Anglo-American Perspective." *Research Policy* 38 (2): 234–242.

Salter, B., and A. Faulkner. 2011. "State Strategies of Governance in Biomedical Innovation: Aligning Conceptual Approaches for Understanding 'Rising Powers' in the Global Context." *Globalization and Health* 7/1: 3–14.

Salter, B., and C. Salter. 2010. "Governing Innovation in the Biomedicine Knowledge Economy: Stem Cell Science in the USA." *Science and Public Policy* 37/2: 87–100.

Sethi, N. 2019. "Regulating for Uncertainty: Bridging Blurred Boundaries in Medical Innovation, Research and Treatment." *Law, Innovation and Technology* 11 (1): 112–133.

SMER. 2016. *Ethical Evaluation at the Interface between Clinical Practice and Research*. Stockholm: Swedish National Council on Medical Ethics (in Swedish.

SOU. 2017. *Ethical Review: An Overview of Regulation of Research and Clinical Practice*. Stockholm: Statens offentliga utredningar (in Swedish.

Stilgoe, J., R. Owen, and P. MacNaghten. 2013. "Developing a Framework for Responsible Innovation." *Research Policy* 42 (9): 1568–1580.

Stirling, A. 2017. "Precaution in the Governance of Technology." In *The Oxford Handbook of Law, Regulation, and Technology*, edited by B. Brownsword, E. Scotford, and K. Yeung, 645–669. Oxford: Oxford University Press.

StratRegen. 2016. "StratRegen." Accessed 20 June 2016. www.stratregen.se

Swedish Executive Government. 2008. "Commission on the Calling For, Selecting, and Evaluating Strategic Research Areas." *Regeringsbeslut* 11: 14. (in Swedish).

Swedish Research Council. 2012. "Follow-up on Strategic Research Areas 2011 - Appendix 2." Accessed 9 June 2016. https://www.vr.se/download/18.4f8b5288136d26da50c1fd/1340207577112/SFOrapport2011+Bilaga+2.pdf

Swedish Research Council. 2015. *Evaluation of the Strategic Research Area Initiative 2010-2014*. Stockholm: Swedish Research Council.

Schwartz, J.A.T. 2014. "Innovation in pediatric surgery: The surgical innovation continuum and the ETHICAL model." *Journal of Pediatric Surgery*. 49: 639–645.

Thomke, S., E. von Hippel, and R. Franke. 1998. "Modes of Experimentation: An Innovation Process—and Competitive—variable." *Research Policy* 27: 315–332.

Thune, T., and A. Mina. 2016. "Hospitals as Innovators in the Health-care System: A Literature Review and Research Agenda." *Research Policy* 45: 1545–1557.

Vignola-Gagné, E., P. Biegelbauer, and D. Lehner. 2014. "Translational Research: Entrepreneurship, Advocacy and Programmatic Work in the Governance of Biomedical Innovation." In *The Governance of Socio-Technical Systems. Explaining Change*, edited by S. Borrás and J. Edler, 132–158. Cheltenham: Edward Elgar Publishing.

Vincenti, W. G. 1990. *What Engineers Know and How They Know It*. Baltimore: Johns Hopkins University Press.

Vogel, G. 2013. "Trachea Transplants Test the Limits." *Science* 340: 266–268.

WMA. 2008. "Declaration of Helsinki." Accessed 6 October 2019. https://www.wma.net/wp-content/uploads/2018/07/DoH-Oct2008.pdf

Yin, R. 1994. *Case Study Research*. Thousand Oaks: Sage Publications.

Appendix – Overview of investigations

No	Date of initiation	Investigations	Initiated by	Date of report	No of pages
1	Nov 2014	External investigation of formal complaints of scientific misconduct	KI Vice chancellor	May 2015	41
2	Feb 2015	Clinical innovation	Swedish National Council on Medical Ethics (SMER)	Nov 2016	82
3	Feb 2016	Circumstances of operations performed 2011–2013	KUH director	Aug 2016	145
4	Feb 2016	External investigation of how KI has handled the Macchiarini case	KI board	Sep 2016	202
5	Feb 2016	Propose recommendations for clinicians and researchers working at the crossroads between clinical research and healthcare	The Royal Swedish Academy of Sciences and Swedish Society of Medicine	Jun 2016	10
6	Feb-Mar 2016	Scientific misconduct (re-opened)	KI vice chancellor	Multiple statements 2017–2018	-
7	Apr 2016	Additional scientific fraud investigation.	KI vice chancellor	Multiple statements 2017–2018	-
8	Jun 2016	Criminal investigation (public prosecutor)	Health and Social Care Inspectorate (IVO) and The Medical Products Agency (LMV) (independently)	Oct 2017	5
9	Jun 2016	Review of regulation concerning research ethics and the interface between clinical research and clinical practice	Executive government (SOU)	Dec 2017	482
10	Mar 2017	Monitoring KI in Feb 2016 – pending investigation.	Swedish Higher Education Authority (UKÄ)	Nov 2017	32

Exposing three dark sides of social innovation through critical perspectives on resilience

Martin Fougère and Eija Meriläinen

ABSTRACT

In this essay we expose three dark sides of social innovation (SI) by mobilising the concept of resilience. We examine implications for SI from (1) resilience *thinking*, (2) (critical) resilience *studies* and (3) the exceptional *contexts* in which resilience is needed. The first dark side of SI is that SIs lead to *disruptions* likely to cause unintended adverse consequences. The second dark side is that top-down SIs tend to be deployed in the name of vulnerable communities, but in neoliberal ways mainly concerned in making these communities more *productive* for society, at the risk of heightening their marginalisation. The third dark side is that SI *discourse* lends itself too easily to hijackings by powerful actors driving their own interests of capital accumulation while calling for communities to self-organise. We discuss how critical perspectives on resilience help us challenge these dark sides of SI.

1. Introduction

Even though innovation has not always been seen as inherently good (Godin 2012), a widely shared contemporary assumption is to see it as something desirable (Coad et al. 2018), aligned with progress, economic growth and (at least implicitly) societal welfare. In line with this assumption, the contemporary notion of social innovation (SI) combines the positively loaded emphasis on novelty with a more explicitly affirmed purpose of contributing to societal welfare, being generally defined as new ideas that work better than those already existing alternatives in addressing social problems (European Commission 2010; Phills, Deiglmeier, and Miller 2008). In this critical essay, we see SI not only as illustrative of the 'goodness' or 'desirability' of innovation broadly construed, but also as particularly revealing of its dark side(s), and notably of potential negative impacts of innovation policy on society. More specifically, having noticed the proposal for a pairing up of SI and resilience in both academic and policy contexts (Moore and Westley 2011; Westley and Antadze 2010; Westley 2013; Tello-Rozas et al. 2017), we examine how resilience thinking, especially when applied to the post-disaster context in which resilience and social innovation are both argued to be particularly needed, helps in exposing some of the 'darkest sides' of SI.

We take to heart the call for 'unorthodox, interdisciplinary and potentially controversial' or even 'heretical' approaches (Coad et al. 2018) to expose the dark side of innovation and take as our starting point the political-economic context in which we (at least in the overwhelming majority of urban settings in the Global North) live, i.e., a neoliberal and neoliberalizing regime. By 'neoliberal' we mean a system which is driven by the pursuit of economic growth (in GDP terms) through market-based solutions,[1] and in which 'wealth creation' is assumed to eventually trickle down to the economically marginalised communities in society – against all the evidence accumulated since the 1970s, which instead shows that inequalities in both developed and developing countries can only be exacerbated if market-based priorities are not counterbalanced by significant redistributive policies (Stiglitz 2016). By 'neoliberalization' we refer to the concurrent ongoing processes, experienced in many countries around the world, of (1) gradual downsizing of state functions ('roll-back' neoliberalization), (2) redeployment of social policies through 'third-sector' and market-driven social innovations ('roll-out' neoliberalization; see Peck and Tickell 2002), and (3) activation of entrepreneurial drives of individual citizens, responsibilized for their own risks ('roll-with-it' neoliberalization; see Keil 2009).

In the Global North, and especially in the Anglo world, both social innovation (SI) and resilience have become influential contemporary buzzwords in the context of social systems and policy (Pol and Ville 2009; Stumpp 2013). They are increasingly invoked when addressing social problems, which gives them the power of *legitimising* policy interventions of various kinds, and thus their deployment has significant impacts on contemporary societies. Both SI and resilience policy discourses share a number of similarities, including:

(1) the fact that they tend to be deployed for 'mending' effects in the context of crises (Fougère, Segercrantz, and Seeck 2017) and/or disasters (Grove 2014a);
(2) their tendency to accept significant disruptions in social relations (Davoudi 2016; Moore et al. 2012) and to lead to a 'new normal' (Fainstein 2015);
(3) their idealization of the capabilities of civil society and/or communities to entrepreneurially self-organize in order to address the social and environmental challenges they face (Fougère, Segercrantz, and Seeck 2017; Grove 2014b); and
(4) their call for a responsibilization of marginalized individuals and communities for the risks they face (Chandler and Reid 2016; Swyngedouw 2009).

It is thus not very surprising that some mainstream SI scholars claim that pairing resilience with SI will be a win-win (Moore and Westley 2011; Westley and Antadze 2010). In SI policy, as in innovation policy at large, the unintended adverse impacts of innovations tend to be comparatively overlooked (Giddens 1999; Mulgan 2016). Complementing SI with systemic resilience thinking then, it is argued, will help anticipate those adverse impacts appropriately (Westley 2013). In more recent years, there have also been calls for combining resilience thinking with SI with the intention of challenging neoliberal hegemony (see Tello-Rozas et al. 2017), notably in the Global South. Indeed, while SI policy discourse tends to privilege 'structural', 'complementary' and 'instrumental' types of SI which do not challenge the predominant economic order (Moulaert et al. 2017), there are also 'radical' types of SI (for a typology, see Marques,

[1]The neoliberal justification for relying only or mainly on market-based solutions comes from the Hayekian notion that the market, as 'an information processor more powerful than any human brain' (Mirowski 2013: 54), knows best what is and is not desirable for society (Hayek 1948).

Morgan, and Richardson 2018) which are meant to address power asymmetries in local sites.

As two authors who have worked on SI and resilience respectively, we take an interest in this suggested pairing up of resilience thinking with SI. Our aim in this essay is to expose the dark sides of contemporary SI by mobilising the concept of resilience, so as to challenge hegemonic understandings of SI. More specifically, we propose to do this by examining in turn some implications for SI from (1) resilience *thinking*, (2) (critical) resilience *studies* and (3) the exceptional *contexts* in which both resilience and SI are argued to be particularly needed. We illustrate the three dark sides of SI through the well-documented case of New Orleans post-Katrina, in many ways an extreme case allowing for looking into the darkest possible sides of SI, in a context of acute shock where drastically 'innovative' measures were taken in the name of 'resilience'.

In the next section we reflect on what is understood, generally speaking, as SI. We then briefly introduce our illustrative case of post-Katrina New Orleans, before moving on to discussing the first dark side of SI, as exposed by resilience *thinking*: SIs cause disruptions, and innovators and policymakers need to very carefully think about possible unintended adverse consequences of their SIs. Looking into critical *studies* of resilience then leads us to exposing a second dark side of SI: that top-down SIs tend to be deployed in the name of 'vulnerable' communities, but in ways that are mainly concerned with making these communities more productive for society, at the risk of making them even more marginalised. Our subsequent section on the exceptional resilience *contexts* where both resilience and SIs are invoked helps us expose the third dark side of SI: that it discursively lends itself to hijackings by powerful actors who drive their own interests of capital accumulation all while wrapping their 'new ideas' in a positive rhetoric emphasising how communities should be empowered to self-organise. In our concluding discussion, we summarise the three dark sides in relation to the different 'powers' of SI and recapitulate key insights from the resilience perspective in post-disaster contexts. Furthermore, it is in the concluding discussion that we more extensively explore to what extent the dark sides also hold for 'innovation' broadly construed, drawing some connections to seminal works from innovation studies (such as Christensen 1997; Lundvall 1992; Nelson 1993; Porter 1990; Rogers 1983; Schumpeter 1942).

2. Framing social innovation

Below are two influential definitions of SI, whose main characteristics can be traced not only in 'Anglo-American' academic and policy papers but also in a wide range of additional policy contexts, such as the European Union (see Fougère, Segercrantz, and Seeck 2017).

> [A social innovation is] a novel solution to a social problem that is more effective, efficient, sustainable, or just than existing solutions and for which the value created accrues primarily to society as a whole rather than private individuals. (*Rediscovering Social Innovation*, Phills, Deiglmeier, and Miller 2008)

> Social innovation is about new ideas that work to address pressing unmet needs. We simply describe it as innovations that are both social in their ends and in their means. Social innovations are new ideas (products, services and models) that simultaneously meet social needs (more effectively than alternatives) and create new social relationships or collaborations. (*This is European Social Innovation*, European Commission 2010: 9)

The main characteristics of SIs, which can also be found in most other definitions in one form or another, relate to: (1) the diagnosis of a 'social problem' (Phills, Deiglmeier, and Miller 2008) or 'pressing unmet needs' (European Commission 2010) to be addressed through SI – i.e. SI needs to be warranted by an argument of existing social or societal problems or needs; (2) the proposal of a 'novel solution' (Phills, Deiglmeier, and Miller 2008) to the problem or of 'new ideas that work' (European Commission 2010) in meeting the pressing needs; and (3) the legitimation of the SI through a claim that it is more effective than 'existing solutions' (Phills, Deiglmeier, and Miller 2008) or 'alternatives' (European Commission 2010), which also entails a negative appraisal of the effectiveness of those existing solutions and alternatives. Following this policy and academic perspective, SI is good by definition, as it is that which is more effective than alternatives. An important consequence of this inherent 'goodness' of SI is that when a new solution, supported by powerful actors, is framed as SI, it is very difficult for those less powerful actors who are affected negatively by the SI to challenge it successfully. While there are many ways to categorise SIs, here we characterise two types of SI based on what or who the target of the initiatives is, and what or who is in need of an SI intervention:

(a) ***SI for 'Vulnerable Communities'***. This type of SI targets populations that are, from the perspective of the innovators, failing to integrate in, cope with or be a productive part of (the ideal) society – this corresponds to what Marques, Morgan, and Richardson (2018) call *targeted* types of SI. Here the vulnerable communities are often both the targets (SI is expected to bring about change in them and for them) and the ones who will be adopting the innovation. While in the context of a natural hazard 'vulnerable community' can mean a community (i.e. 'an entity that has geographic boundaries and shared fate', as in Norris et al. 2008) that is exposed to harm, sensitive and lacking adaptive capacity (Smit and Wandel 2006; Yarnal 2007), here we understand the concept more broadly. A 'vulnerable community' can for example consist of young students needing to exercise more because of an overall posited obesity risk, or of unemployed people deemed to be unproductive and in need of re-education to become employable again, or of contemporary citizens with an 'app' or 'smart phone' addiction problem.

(b) ***SI for Society***. While SIs might be targeted to certain 'vulnerable communities', they may also be intended to address issues that are plaguing society as a whole – this corresponds to what Marques, Morgan, and Richardson (2018) call *structural* types of SI. This type of SI may still be mainly targeting 'vulnerable communities' when it tries to integrate a certain population 'better' in the existing or ideal shape of society. However, when SI aims to alter the structures of society as a whole – for example through altering how healthcare is provided for all citizens, or through extreme tax breaks with far-reaching consequences for public services – it can be labelled 'SI for society'. Marques, Morgan, and Richardson (2018) claim that structural SI has become less relevant as an analytical construct since social scientists have stopped focusing on structural analysis of large transformations, but it is alive and well in policy discourse, where SI is called upon to deliver the structural transformation that is needed in society (see Fougère, Segercrantz, and Seeck 2017). This type of SI is particularly prevalent in areas or moments that experience major shocks or disasters, as some elite neoliberal actors attempt to seize the opportunity provided by the massive disruption to significantly transform society towards further neoliberalization (Klein 2007). In more general terms, it is relevant wherever neoliberalization is occurring, not only in developing areas subjected to structural adjustment programmes but also in seemingly milder neoliberalization processes, including in continental Europe (Fougère, Segercrantz, and Seeck 2017).

Both types of SI above entail forms of *intervention*: either to change the trajectories of the 'vulnerable communities', or to change – or reinforce – the political course of society. Thus, any SI is likely to cause *disruptions*, either for the 'vulnerable communities' or for society as a whole – and often for both, as the already marginalised communities may be more exposed to the adverse consequences of societal disruptions. It is this disruption and its systemic ripples, the intended and unintended consequences of the SI, that resilience thinking could be hoped to shape in a favourable manner (see Westley 2013). However, while any powerful SI will lead to a significant disruption, it should be noted that SI interventions also tend to be deployed at times when a disruption to a community's or society's everyday has already happened. For example, a natural hazard can disrupt the lives of marginalised people, leading for them to be framed even more as a vulnerable community – and thus subjected to SI (as in (a) above). Or otherwise, a natural hazard can act as a catalyst to push neoliberal policies further in society by deploying 'new ideas that work' based on a certain construction of the problem and the alternative solutions (as in (b)). In the next section we briefly introduce the illustration of New Orleans post-Katrina, before moving on to discussing in more detail three ways in which studies of resilience help us expose the dark side(s) of SI, by looking in turn into (1) resilience thinking, (2) critical studies of resilience, and (3) exceptional contexts in which resilience and SI are both argued to be needed.

3. Social innovation and resilience after hurricane katrina: New Orleans as illustration

As an illustration of the three dark sides of social innovation, we will refer to the extensively documented and studied case of New Orleans post-Katrina, inspired by a number of academic accounts on the case (Akers 2012; Barrios 2011; Cupples and Glyn 2014; Davies 2017; Elliott and Pais 2006; Green, Bates, and Smyth 2007; Paidakaki and Moulaert 2018; Peck 2006; Rowe 2014; Stephens et al. 2009; Thiele 2017; Yarnal 2007; Zebrowski and Sage 2017). We mobilise this case because in the aftermath of Hurricane Katrina in the short-term as well as in the long-term, New Orleans was a site where (1) the concept of resilience was particularly invoked (Paidakaki and Moulaert 2018; Rowe 2014; Zebrowski and Sage 2017), and (2) many different types of SIs were launched, with a number of community-based innovation initiatives competing with neoliberal initiatives (Paidakaki and Moulaert 2018; Rowe 2014) and innovative top-down policy moves (Barrios 2011; Zebrowski and Sage 2017). In her work, Mary Rowe, who founded the New Orleans Institute for Resilience and Innovation, makes a very explicit link between resilience and community-based innovation and introduces a number of inspirational examples of community-based innovation 'hubs' that contributed to resilience in the city (Rowe 2014). But SIs for resilience were also part of a major policy effort by the City of New Orleans, which, in its document *Resilient New Orleans* (City of New Orleans 2015), describes resilience as a matter of investing in 'equity', 'a term which deliberately blurs the distinction between pursuing social equality and positioning communities as an investment opportunity' (Zebrowski and Sage 2017, 50).

Paying attention to both top-down and bottom-up SIs in post-Katrina New Orleans, Paidakaki and Moulaert (2018) show how many pro-growth actors (real estate developers and companies) attempted to seize the opportunity to strengthen their hold over the city by appealing to a hegemonic, neoliberal resilience approach. They also document a number of counter-hegemonic 'social resilience cells' (SRCs) which emerged as a resistance and response to both the neoliberal hegemony itself and the post-disaster needs of communities and neighbourhoods in New Orleans (ibid.). According to Paidakaki and Moulaert (2018, 438), these counter-hegemonic SIs were of two main kinds: 'pro-equity' and 'pro-comaterializing'. Pro-equity SRCs work with various civil society and public actors (including progressive foundations, charity groups, and/or state agencies) to advocate use value rather than exchange value of houses and a fairer distribution of financial resources across housing providers – with the objective of making access to existing housing easier for marginalised communities. The pro-comaterializing SRCs are focused on concretely building houses based on a model of solidarity and grassroots involvement – with the objective of working together with marginalised communities in building their own housing. Thus, it is in a contested terrain between pro-growth actions and more socially minded initiatives that New Orleans's SIs for resilience were developed.

4. Resilience thinking: on the disruptive power of social innovation

Resilience as a concept and way of thinking has developed within a variety of disciplines such as mechanics, psychology and ecology, and more recently resilience has become a buzzword in the fields of disaster risk reduction, sustainability science and climate change adaptation (Alexander 2013). The core idea is that resilient systems can, after shocks, either return to a state of normalcy or equilibrium (the original understanding of resilience) or at least absorb changes in such a way that the systems persist over time (Holling 1973). Within disaster risk reduction, for example, resilience thinking is sometimes used to posit that communities or societies can recover from societal disruptions with little external assistance (Manyena 2006). Resilience is often criticised for being a conservative concept in that it strives back to a status quo that may have been undesirable to begin with (MacKinnon and Derickson 2013). Some authors have deployed 'improved' versions of resilience, such as evolutionary resilience (White and O'Hare 2014), or they have coupled resilience with adaptation, resistance, incremental adjustment and/or transformation (Matyas and Pelling 2015; Smit and Wandel 2006).

Resilience thinking can invite innovators and policymakers to see SI not merely as a win-win(-win) solution or panacea, but also as an additional cause of disruption with possible unintended adverse consequences on other parts of a social system. This perspective stemming from resilience thinking would be desirable in the light of Fougère et al. (2017: 840) conclusions that state that

> for SIs to be as progressive as possible, it is crucial that [their] win-win mindset be (1) always deployed in context, in terms of very specific concrete conditions, (2) not prioritizing the economic win over social wins, and (3) not blinding the innovators to possible negative

impacts of their SIs, whether directly or through causing the displacement of previous social benefits (by contributing to further neoliberalization).

Resilience thinking can invite social innovators to think more holistically about overall system effects of their new ideas that may disrupt and/or displace existing social benefits.

Even based on the more recent, evolutionary perspective on resilience, which does not necessarily entail a return to normality after a shock (see Davoudi 2016), social innovators would need not only to think of how their solutions work well to solve specific problems in isolation as it were, but also to consider the broader system impacts the solutions might have (Westley 2013). This means that the transformative nature of SI, which entails a disruptive dimension, demands that innovators assess the broader impacts of the disruption they cause. The disruptive power of SI is not per se a problem – in fact all innovation might imply some degree of disruptive change – but with such disruptive power comes a potential dark side, which resilience thinking calls for evaluating carefully.

Under certain conditions of disruption, the concepts of SI and resilience could be complementary. Following a disruption (due to a natural hazard for example), 'resilience' would strive to stabilise the system, while SIs designed to change the trajectory of either a 'vulnerable community' or the whole society may be initiated to complement the resilience approach. While there is potential in this coupling, there are also severe pitfalls with this approach, not least because SIs do cause disruptions too. This is indeed the first exposed 'dark side' of SI, i.e. by seeking to address societal problems through novel solutions, SI might cause problematic disruption, thereby creating new problems while not prioritising the attempt to reclaim what has been lost.

The case of New Orleans illustrates these issues insightfully: people and communities lost, in the first instance, their housing – many of the most marginalised people had to move away from New Orleans, without a possibility to ever return (Elliott and Pais 2006). Many also lost their livelihoods and what gave meaning to their lives, and a transformational thinking in terms of SIs favoured a *replacement* of the most marginalised people by less marginalised people rather than an opportunity for marginalised people to reclaim what they lost (Barrios 2011; Peck 2006; Zebrowski and Sage 2017). When it comes to what used to give meaning to people's lives in various neighbourhoods of New Orleans, Barrios (2011) for example discusses how music has long held special significance in Tremé, where going back to music traditions (including parades) was perceived as a key priority by the community. In a similar vein, Thiele (2017) looks into how important local religious beliefs and rituals were for some of the marginalised communities to be able to cope with the disaster.[2]

5. Critical resilience studies: on the productive power of social innovation

In contrast with theoretical resilience *thinking*, critical resilience *studies* suggest that many resilience initiatives striving to address the impacts of shocks tend to expect resilience and adaptability from those facing a disruption, rather than facilitating and

[2]The TV-series *Tremé* (Simon and Overmyer 2010-2013), partially based on ethnographic accounts, shows how difficult it was for African American communities to be able to return to the practices and perform the rituals that could help them to cope in the aftermath of Katrina.

building resilience on a systemic level. Resilience as a largely top-down, expert discourse may serve to conceal the material processes stemming from capitalist social relations that shape the marginalisation of people made and framed as 'vulnerable' (MacKinnon and Derickson 2013; Bankoff 2019). Rather than addressing the production of vulnerability and the 'social problem' as it were, the mainstream resilience approach instead tends to imply that the communities are the problem and that they need to change through mobilising their immanent capacities for self-organised adaptation (Grove 2014b; Meriläinen 2020). Grove (2014a; 2014b) unpacks resilience discourse as a neoliberal project of social engineering which makes the vulnerability and resilience of communities 'an object of governmental intervention and control' (Grove 2014a, 240), under the guise of empowering marginalised groups to manage their own vulnerabilities and integrate better in society.

In a similar way as with resilience policy discourse, SIs typically are framed as attempts to help vulnerable communities. Yet in order to work, many SIs rely on the immanent capacities of these vulnerable communities, asking them to transform dramatically, in ways that are not asked of 'non-vulnerable' people, and possibly making them even more marginalised in this process. For example, an SI meant to make long-term unemployed people employable again might force them (by conditioning unemployment benefit) to drastically re-train themselves towards a completely different occupation, requiring them to activate themselves, responsibilize themselves for the risks they face, and become more enterprising. In SI policy discourse, this is what makes it possible to argue not only that SI addresses social problems, but also, much like innovation more broadly construed, that it makes society more productive. This is a second dark side of SI, as practiced top-down by innovators and policymakers: in the name of helping the vulnerable (e.g. in the name of health, or inclusion), it often entails putting the capacities of the marginalised under heightened stress while doing little to address structural reasons for unemployment (or other social issues), thereby often worsening vulnerabilities. In the example above, those who fail to become employable can only blame themselves if they end up being even more marginalised (cf. Chandler and Reid 2016), while indicators will show that society has been made more productive *and inclusive* through the successful enhanced employability of some of the targeted people.

In the context of New Orleans post-Katrina, Barrios (2011, 120) documents how in a so-called participatory planning process, the innovative plan for a 'Unified New Orleans' (UNOP) was driven by expert planners in line with neoliberal principles of urban development which 'propose the evaluation of all aspects of social life under rubrics of capitalist utility and cost-benefit'. Drawing in part on Barrios's (2011) work, Zebrowski and Sage (2017, 51) note how these expert urban planners viewed resilience as a matter of 'economic productivity, entrepreneurship and competition', and neighbourhoods as sites of capital investment, not as meaningful places for communities, so much so that in this form of urban planning, 'the loss of a section of the population [...] is only regarded as problematic if [it] negatively impacts on measures of urban competitiveness or productivity' (ibid.). Thus, the disruptive vision of a plan like UNOP, seeking to transform neighbourhoods to make them more productive, was clearly not meant to help the most marginalised people, who had lost everything after Katrina and could not go back home. Here, both the disruptive and productive powers of SI reveal their dark sides: rather than alleviating the suffering of the most marginalised communities, UNOP contributed to replacing these communities in order to 'upgrade' the poor neighbourhoods.

Another insight from critical resilience studies is that while neoliberal resilience discourse attempts to govern people through their immanent adaptive and productive capacities, the very reliance on these capacities means that it can be subverted from a bottom-up perspective (Grove 2014b). MacKinnon and Derickson (2013) suggest that we should resist the use of 'resilience' as a term and employ an alternative signifier – such as 'resourcefulness' – as a basis for a bottom-up, counter-hegemonic discourse. Similarly, marginalised communities, through SIs that are variously called 'bottom-up', 'grassroots' and/or 'community-based' (Rowe 2014), might be able to address the causes of the social problems through using their immanent capacities to challenge neoliberal hegemony through 'alternative organizing' (Zebrowski and Sage 2017), instead of either (1) obediently providing the economically productive subjects that this hegemony demands or more likely (2) falling further down in a vulnerability spiral.

Thus, key insights of critical resilience *studies* for understanding the dark sides of SI relate to the dynamics of top-down vs. bottom-up initiatives, and more specifically how marginalised communities are likely to be better off when being innovators themselves rather than when being framed as vulnerable 'beneficiaries' of top-down SIs.

6. Exceptional resilience contexts: on the discursive power of social innovation

The theoretical insights reached within critical resilience studies can be complemented with illustrations from those exceptional *contexts* where resilience and SI are typically called for, that is, in areas and communities prone to disasters – and especially in the aftermath of disasters. In this regard, the extreme case of New Orleans post-Katrina exposes a third dark side of SI. Because of SI's framing as 'good' – that is, 'better than existing alternatives' – by definition, the concept lends itself to being used by elite actors, under the guise of 'addressing social problems', but in order to advance their own interests, often at the cost of marginalised communities and society as a whole. It has been noted by many authors (not least Klein 2007) that crises and disasters have been historically seized as opportunities for accelerated neoliberalization, whether globally or more locally. In the case of New Orleans post-Katrina, many elite interventions were framed exactly along the lines of how SI is typically defined in Anglo-American literature – as per the main characteristics in the two influential definitions cited in the beginning of the article (European Commission 2010; Phills, Deiglmeier, and Miller 2008). These defining characteristics of SIs suggest how SIs can be framed by powerful actors. Namely, actors with the power to define the 'social problem' (Phills, Deiglmeier, and Miller 2008) and/or the 'pressing unmet needs' (European Commission 2010) may be tempted to do this in terms of the types of 'novel solution' (Phills, Deiglmeier, and Miller 2008) and/or 'new ideas' (European Commission 2010) that are in their interest and that they can deem 'more effective' (Phills, Deiglmeier, and Miller 2008; European Commission 2010) than existing solutions or alternatives. This entails discrediting the latter alternatives as old ideas that do not work.

Thus, in New Orleans post-Katrina, a number of policy influencers leveraged SI discourse by (1) deploying their power to define the social problem, (2) very strongly driving their own 'effective' solutions while also actively (3) discrediting 'less effective' alternatives. Peck's (2006) detailed account of the role of certain think tanks immediately after Katrina makes it possible to discuss these three points in turn.

6.1. The power to define the social problem

The social problem that needed addressing through new SIs in New Orleans was actively and systematically discursively constructed by neoliberal think tanks right after Katrina, with the claim that the hurricane 'exposed a decadently corrupt, welfare-dependent and murderously uncontrolled city' (Peck 2006: 684). Manhattan Institute analyst Nicole Gelinas wrote multiple pieces in different outlets representing New Orleans as a completely failed city:

> New Orleans can't take care of itself even when it is not 80 percent underwater ... The city's government has long suffered from incompetence and corruption ... New Orleans teems with crime, and the NOPD can't keep order on a good day .... Socially, New Orleans is one of America's last helpless cities. (Gelinas 2005c, 9, in Peck 2006, 693)

> Katrina didn't turn innocent citizens into desperate criminals. This week's looters ... are the same depraved individuals who have pushed New Orleans' murder rate to several multiples above the national average in normal times ... On a normal day, those who make up New Orleans' dangerous criminal class – yes, likely the same African-Americans we see looting now – terrorize their own communities ... Failure to put violent criminals behind bars in peacetime has led to chaos in disaster ... The city must be forcefully demilitarized, even as innocent victims literally starve. (Gelinas 2005a, 2–4)

Gelinas's colleague at the Manhattan Institute, Kay Himowitz, made it clear, in the same *City Journal*, that 'it was now down to conservatives to respond to the real problem: the breakdown of the nuclear family, together with its racially and socially uneven consequences' (Peck 2006, 702). In the same vein, Gelinas (2005b, 22) identified the core problem as boiling down to 'New Orleans' legions of weak, femaleheaded, underclass Black families supply generation after generation of ... 'lightly parented' young men to fuel the carnage'. The American Enterprise Institute, another powerful neoliberal think tank, included in its monthly magazine *The American Enterprise* similar condemning diagnoses of New Orleans's pressing problem, blaming the 'rotten administration' of urban welfarism for a 'collapse of responsibility and discipline' (Peck 2006, 703). Joel Kotkin thus wrote that

> If we are lucky, the flood waters of Katrina will wash away some of the '60s-era illusions that fed today's dysfunction. Honest observers will recognize that this natural disaster, which hit the nation so hard, was set up by the man-made disaster of a counterproductive welfare state. (Kotkin 2006, 29, in Peck 2006, 703-704)

George Neumayr, an editor of the conservative monthly *American Spectator*, added that 'the chaos after Hurricane Katrina did not cause a civilizational collapse; it simply exposed and magnified one that had already occurred' (Neumayr 2005, 48, in Peck 2006, 704). Thus, the main problems that needed tackling in the aftermath of Katrina were framed by neoliberal think tanks to be, not the safe return of those who had lost their homes, but instead issues of 'welfare dependency, crime, family breakdown and corruption' (Peck 2006, 705). The emphasis on welfare made sure that the government was considered part of the problem, as was, by extension, 'the enormous financial cost of the recovery effort' (Peck 2006, 706), which was deemed a problem in itself as it embodied a lack of fiscal discipline that needed to be fought through budget reductions. This eventually ensured that the worst victims of Katrina, the economically and socially marginalised, were least helped by the recovery effort. Not only were they not helped, they were also considered part of the problem, as their need for support was framed as undeserved entitlement and a lack of individual responsibility. Once the problem had been framed in a way conducive to neoliberal reform, neoliberal 'solutions' and 'new ideas' could be brought forward.

6.2. *The power to impose the 'new ideas' and the 'solutions'*

Peck (2006) documents how very quickly after Katrina, neoliberal think tanks constructed

> a 'principled response' ... predicated upon fiscal restraint and 'offsetting' budget cuts in Washington, DC ... enlarging the role for private enterprise through market-led development, governmental outsourcing, and city governance; selective institutional roll-backs focused on the social state; redoubled crime control, making the city safe for tourists and gentrifiers; and an interventionist program of 'moral reconstruction' aimed at those stranded in the storm's wake. (Peck 2006, 692-693)

In pure shock doctrine style, less than two weeks after Katrina, the Heritage Foundation already had come up with its full package of principled solutions: this involved delimiting the minimal involvement of the federal government, tax credits and voucher programmes to encourage private-sector innovation, reduction of red tape to accelerate private investment in rebuilding efforts, a systematic shifting of resources from low-priority to high-priority uses, and presenting private entrepreneurship as the only needed engine for rebuilding efforts (Meese, Butler, and Holmes 2005). In particular, Meese, Butler, and Holmes (2005) called for declaring New Orleans and other affected areas 'Opportunity Zones', for which an Emergency Board led by the President should 'identify regulations at all levels that impede recovery and should propose temporary suspension or modification of these rules'. Another keyword in these principled solutions was 'offsets', meaning that any higher spending from government would need to be offset by reductions elsewhere. All these suggested measures, of course, were presented as forms of SI, as ways to 'experiment boldly' (Rector 2005, in Peck 2006: 702), that is, seize the opportunity provided by the disaster to accelerate neoliberal 'programmes of 'social regression and market governance' (Peck 2006, 708), in order to advance further capital accumulation for private investors, while discrediting the state as much as possible.

6.3. *The power to discredit the 'less effective' old ideas*

In contemporary SI discourse, those 'alternatives' that are deemed 'less effective' than the new ideas tend to be publicly funded and welfare-related. In New Orleans, the construction of the social problems and the neoliberal recommendations for solutions were complemented by a systematic effort to discredit state efforts. Part of this disqualification of state legitimacy occurred through the critique of welfarism as a key part of the problem in New Orleans, but in addition, an SI-inspired discourse coloured the critique of the state, as state action was systematically deemed to be problematic because of its 'top-down' nature, contrasted with the potential for a libertarian 'laboratory for ideas' that a New Orleans freed from public spending would have the potential to be (Glassman 2005, 2, in Peck 2006, 698). As Feulner put it:

> [We] shouldn't respond to government failures by making the government larger and still more unwieldy. The best way to rebuild New Orleans will be for the government to get out of the way. Congress and state governments can do this by eliminating or reducing regulations and allowing communities to decide for themselves how best to rebuild. (Feulner 2005)

The bottom-up self-organising of communities against top-down state intervention is here used by a Heritage Foundation writer, and this is one of the ways in which self-organising-oriented discourses like community resilience and SI can be co-opted by neoliberal elites to promote social regression for marginalised communities. Similarly, powerful think tanks often present reforms that have been adopted as a result of their own strategic initiatives as 'an outcome of grassroots energy' (Peck 2006, 700). Thus, these think tanks leveraged the discursive power of SI, that is, they used the positivity of SI discourse – its connotations of effectiveness, equity, grassroots initiation, inclusion, etc. – to legitimise their neoliberalizing interventions.

7. Concluding discussion: with great powers ... come great dark sides

In this essay we have examined dark sides of SI by drawing on insights from (1) resilience *thinking*, (2) (critical) resilience *studies* and (3) exceptional resilience *contexts*. Each of these three perspectives on resilience has helped us to uncover one dark side of SI. In exposing these three dark sides of SI, we have striven to show that each of them is the flip side of a power of SI that enables SI's promise of progressive transformation. Table 1 summarises the three dark sides of SI, each named after its connection to a specific 'power' of SI. In the table, the dark sides are characterised based on: (1) the extent to which they lie in the nature of SI versus the discourse on SI; (2) the extent to which they also hold for 'innovation' broadly construed (hereafter referred to simply as 'innovation'); (3) illustrations from the case of New Orleans post-Katrina; and (4) insights from critical perspectives on resilience.

The first dark side of SI, connected to its disruptive power, relates to the nature of SI as something that is meant to bring about some change in social relations, at some level. Here, the difference between SI and innovation is not significant. All forms of innovation entail the *implementation* of something *new* (Marques, Morgan, and Richardson 2018), resulting in a change with certain impacts on social relations, a disruption the consequences of which may vary from very minor to world-changing. This disruptive power has been at the heart of innovation studies for a long time, going back to at least Schumpeter's (1942) seminal notion of 'creative destruction', and with prominent examples in contemporary innovation scholarship, such as Christensen's (1997) influential work on 'disruptive innovation'. The problematisation of the disruptive dark side of innovation then, takes us back to another approach that used to be very influential in innovation studies, Rogers's sociological perspective on the diffusion of innovations. Rogers (1983) studied societal consequences, including possible negative impacts of disruptions, as an important aspect of innovations. In relation to this first dark side of SI, important insights from critical perspectives on resilience include (1) the injunction to

Table 1. Three dark sides of social innovation and insights from critical perspectives on resilience.

	1. Disruptive dark side of SI	2. Productive dark side of SI	3. Discursive dark side of SI
Nature of vs. discourse on SI	Nature of SI	Nature of and discourse on SI	Discourse on SI
Relevance to innovation broadly construed	Highly relevant; disruption the most defining feature of both innovation and SI	Moderately relevant; what makes SI productive is different from what makes innovation productive	Relevant; while innovation is not as strongly positive as ethically loaded SI, legitimation through innovation is common in Anglo policy contexts
Illustrations, including in New Orleans	SIs that transform rather than help people reclaim what they lost (in New Orleans, their housing, livelihood and meaning of life)	SIs that claim to address vulnerabilities by making people more productive (in New Orleans, SIs using entrepreneurship to tackle inequalities)	Policies driving elite interests of capital accumulation and legitimated through appeals to the discursive power of SI (in New Orleans, think-tank driven programmes appealing to SI to make the city a tax-free 'opportunity zone')
Insights from critical perspectives on resilience	Injunction to assess well the impacts of disruptions (e.g. Westley 2013) and focus on regaining what has been lost (e.g. Barrios 2011; Thiele 2017)	Critique of top-down focus on making people productive (e.g. Grove 2014b; Zebrowski and Sage 2017) and promotion of targeted bottom-up initiatives (e.g. Rowe 2014)	Careful analysis of how radical transformations are driven (e.g. Peck 2006) and support for counter-hegemonic social innovations for resilience to challenge exchange value-dominated hegemony (e.g. Paidakaki and Moulaert 2018; Zebrowski and Sage 2017)

assess the impacts of the disruptions on a systemic level (Westley 2013), and (2) in the post-disaster context, a notion that transformation at all cost may in fact prevent people and communities framed as 'vulnerable' from having a chance to regain some of what they lost (Barrios 2011), including the important spiritual beliefs and rituals that would help them to cope (Thiele 2017).

The second dark side of SI, connected to its productive power, is a function of both the *nature* of SI as a phenomenon and the *discourse* on SI. SI as a phenomenon is often about helping marginalised people to become better integrated in society (Marques, Morgan, and Richardson 2018), whether through education, healthcare, or inclusion focusing on employability or entrepreneurship, and the promise that these people will be more productive for society is at least implicit in many SI initiatives. But it is in SI policy *discourse* that the need to align SI with economic objectives of growth, competitiveness, productivity and entrepreneurship is most explicitly emphasised, given the priority to economic measurements of societal impacts in contemporary policy in the Global North.

Here, there is a significant difference between innovation and SI. As Marques, Morgan, and Richardson (2018) note, technological innovation by definition is something that is productive, since according to the OECD (2005) hugely influential Oslo Manual, a new product or idea will be called an innovation only if it delivers value for a business. Meanwhile for SI, the purpose is *supposed* to be about responding to certain societal needs, which means that typically 'profit will not be [the] primary goal' (Marques, Morgan, and Richardson 2018, 500). In SI policy discourse, however, much is made of the potential for SI to enhance competitiveness of entire societies, all while addressing the most important social problems, in a claimed win-win (see Fougère, Segercrantz, and Seeck 2017). The corresponding dark side of SI, then, lies in its *policy* preference for addressing people's vulnerabilities by activating and responsibilizing them for their own risks, with the assumption that if they can be empowered to become entrepreneurial or at least employable, it will be a social-economic win-win. What is left invisible in this win-win narrative is the possible negative externalities of projects measuring success in economic terms; for example, projects to develop New Orleans neighbourhoods by attracting more entrepreneurial people and (perhaps unwittingly) making it impossible for marginalised former inhabitants to return.

In relation to the productive dark side of SI, important insights from critical perspectives on resilience include (1) a critique of the top-down focus on making people productive (Barrios 2011; Grove 2014b; Zebrowski and Sage 2017) and (2) the subversion of such productive top-down SIs through the promotion of targeted bottom-up SIs (Rowe 2014) and alternative organising (Zebrowski and Sage 2017). While the productive power of SI is not as straightforward a matter as the productive power of innovation broadly construed, the productive dark side of SI still holds some relevance in the more general context of innovation policy discourse. The academic and policy framing around 'national innovation systems' (Lundvall 1992; Nelson 1993), premised on the idea that we should think in terms of national competitiveness and competitive advantages (Porter 1990), has been so influential that it may have contributed to an uncritical notion that more innovation is always good for society (because it supports economic growth, competitiveness and/or productivity). Such uncritical thinking can be challenged by drawing attention to negative externalities, such as adverse environmental impacts and other possible undesirable consequences of innovations (see Sveiby, Gripenberg, and Segercrantz 2012).

Finally, the third dark side of SI is only a function of SI discourse (mainly policy discourse), connected to its discursive power to legitimise certain policy interventions. Here too, there are differences between SI and innovation, because SI policy discourse plays *both* on the 'productiveness' of innovation *and* on the notion that it is by definition ethically good for society – more effective than alternatives in addressing social needs. In this sense, there is an excess of 'goodness' in SI discourse, which makes it particularly potent in legitimising policy interventions, as long as they are framed persuasively as (1) addressing the most relevant social problems, (2) through appropriate solutions (3) that are more effective than existing alternatives. Thus, the third dark side of SI lies in the ability of certain powerful actors to leverage SI discourse for their own purposes in certain contexts that make drastic transformation possible. It so happens that those contexts most ripe for drastic transformation are 'shock' contexts (Klein 2007), i.e. exactly those where resilience is particularly needed.

In relation to this third dark side of SI, important insights from critical perspectives on resilience include (1) the need to carefully examine how elite actors leverage pro-innovation and grassroots-sounding discourses to drive their projects of drastic transformation towards less support for the most marginalised (Peck 2006) and (2) the injunction to support counter-hegemonic social innovations for resilience to challenge exchange value-dominated hegemony through alternative forms of bottom-up organising (Paidakaki and Moulaert 2018; Zebrowski and Sage 2017). While the excess of 'goodness' in SI discourse makes this third dark side more striking for SI than for innovation, it remains that contemporary policy tends to be characterised by a general 'pro-innovation bias' (e.g., Abrahamsson 1991; Rogers 1983) and that innovation policy discourse has been powerful at legitimising policy interventions in the name of economic growth and competitiveness. Thus, this third dark side is also relevant for innovation more generally.

Thus, mobilising perspectives related to resilience has allowed us to expose three dark sides of SI which also characterise, to different extents, innovation broadly construed. It is important to acknowledge that most of our observations are particularly based on understandings of SI and resilience developed in the Anglo world and relevant in most of the urban Global North, and that the case of New Orleans post-Katrina should be seen as an extreme case which provides particularly striking illustrations of both (1) the three dark sides of SI in neoliberalizing post-disaster contexts and (2) alternative approaches to resilience and SI that are meant to challenge neoliberal hegemony.

Disclosure statement

No potential conflict of interest was reported by the authors.

ORCID

Martin Fougère http://orcid.org/0000-0002-4780-6832
Eija Meriläinen http://orcid.org/0000-0001-7813-9588

References

Abrahamson, E. 1991. "Managerial Fads and Fashions: The Diffusion and Rejection of Innovations." *Academy of Management Review* 16 (3): 586–612. doi:10.5465/amr.1991.4279484.

Akers, J. M. 2012. "Separate and Unequal: The Consumption of Public Education in Post-Katrina New Orleans." *International Journal of Urban and Regional Research* 36 (1): 29–48. doi:10.1111/ijur.2012.36.issue-1.

Alexander, D. E. 2013. "Resilience and Disaster Risk Reduction: An Etymological Journey." *Natural Hazards and Earth System Sciences* 13 (11): 2707–2716. doi:10.5194/nhess-13-2707-2013.

Bankoff, G. 2019. "Remaking the World in Our Own Image: Vulnerability, Resilience and Adaptation as Historical Discourses." *Disasters* 43 (2): 221–239. doi:10.1111/disa.2019.43.issue-2.

Barrios, R. E. 2011. ""If You Did Not Grow up Here, You Cannot Appreciate Living Here": Neoliberalism, Space-time, and Affect in Post-Katrina Recovery Planning." *Human Organization* 70: 118–127. doi:10.17730/humo.70.2.d4356255x771r663.

Chandler, D., and J. Reid. 2016. *The Neoliberal Subject: Resilience, Adaptation and Vulnerability.* London: Pickering & Chatto Publishers.

Christensen, C. 1997. *The Innovator's Dilemma.* Boston: Harvard Business School Press.

City of New Orleans. 2015. *Resilient New Orleans: Strategic Actions to Shape Our Future City.* New Orleans: City of New Orleans.

Coad, A., P. Nightingale, J. Stilgoe, and A. Vezzani. 2018. "The Dark Side of Innovation. Special Issue Call for Papers." *Industry and Innovation.*

Cupples, J., and K. Glynn. 2014. "The Mediation and Remediation of Disaster: Hurricanes Katrina and Felix In/and the New Media Environment." *Antipode* 46 (2): 359–381. doi:10.1111/anti.v46.2.

Davies, D. 2017. "Postcolonial Politics and the Affect Economy: Resisting and Perpetuating Violence through Graphic Representations of Hurricane Katrina." Paper presented at the RGS-IBG Annual International Conference, London.

Davoudi, S. 2016. "Resilience and Governmentality of Unknowns." In *Governmentality after Neoliberalism*, edited by Mark Bevir, 210–249. New York, NY: Routledge.

Elliott, J. R., and J. Pais. 2006. "Race, Class, and Hurricane Katrina: Social Differences in Human Responses to Disaster." *Social Science Research* 35 (2): 295–321. doi:10.1016/j.ssresearch.2006.02.003.

European Commission 2010. "This Is European Social Innovation." European Union. http://ec.europa.eu/DocsRoom/documents/19042

Fainstein, S. 2015. "Resilience and Justice." *International Journal of Urban and Regional Research* 39 (1): 157–167. doi:10.1111/ijur.v39.1.

Feulner, E. J. 2005. "Don't Bind New Orleans in Red Tape." Heritage Foundation. https://www.heritage.org/government-regulation/commentary/dont-bind-new-orleans-red-tape

Fougère, M., B. Segercrantz, and H. Seeck. 2017. "A Critical Reading of the European Union's Social Innovation Policy discourse:(Re) Legitimizing Neoliberalism." *Organization* 24 (6): 819–843. doi:10.1177/1350508416685171.

Gelinas, N. 2005a. "A Perfect Storm of Lawlessness." *City Journal*, September 1, 1–4.

Gelinas, N. 2005b. "Who's Killing New Orleans?" *City Journal* 15: 14–27.

Gelinas, N. 2005c. "Will New Orleans Recover?" New York Sun, September 1, 9.

Giddens, A. 1999. "Risk and Responsibility." *The Modern Law Review* 62 (1): 1–10. doi:10.1111/mlr.1999.62.issue-1.

Glassman, J. K. 2005. "How to Rebuild a Great City." *Scripps Howard News Service*, September 12, 1–2.

Godin, B. 2012. "Social Innovation: Utopias of Innovation from C. 1830 to the Present." *Project on the Intellectual History of Innovation* 11: 1–52. http://www.csiic.ca

Green, R., L. K. Bates, and A. Smyth. 2007. "Impediments to Recovery in New Orleans' Upper and Lower Ninth Ward: One Year after Hurricane Katrina." *Disasters* 31 (4): 311–335. doi:10.1111/j.1467-7717.2007.01011.x.

Grove, K. 2014a. "Agency, Affect, and the Immunological Politics of Disaster Resilience." *Environment and Planning D: Society and Space* 32 (2): 240–256. doi:10.1068/d4813.

Grove, K. J. 2014b. "Adaptation Machines and the Parasitic Politics of Life in Jamaican Disaster Resilience." *Antipode* 46 (3): 611–628. doi:10.1111/anti.v46.3.

Hayek, F. 1948. *Individualism and Economic Order*. Chicago, IL: University of Chicago Press.

Holling, C. S. 1973. "Resilience and Stability of Ecological Systems." *Annual Review of Ecology and Systematics* 4 (1): 1–23. doi:10.1146/annurev.es.04.110173.000245.

Keil, R. 2009. "The Urban Politics of Roll-with-it Neoliberalization." *City* 13 (2–3): 230–245. doi:10.1080/13604810902986848.

Klein, N. 2007. *The Shock Doctrine: The Rise of Disaster Capitalism*. London: Macmillan.

Kotkin, J. (2006). "Ideological Hurricane." *The American Enterprise*, January/ February, 32 24–29. 10.1016/j.jhsb.2006.09.007.

Lundvall, B.-Å. 1992. *National Systems of Innovation – Towards a Theory of Innovation and Interactive Learning*. London: Pinter.

MacKinnon, D., and K. D. Derickson. 2013. "From Resilience to Resourcefulness: A Critique of Resilience Policy and Activism." *Progress in Human Geography* 37 (2): 253–270. doi:10.1177/0309132512454775.

Manyena, S. B. 2006. "The Concept of Resilience Revisited." *Disasters* 30 (4): 434–450. doi:10.1111/j.0361-3666.2006.00331.x.

Marques, P., K. Morgan, and R. Richardson. 2018. "Social Innovation in Question: The Theoretical and Practical Implications of a Contested Concept." *Environment and Planning C: Politics and Space* 36 (3): 496–512.

Matyas, D., and M. Pelling. 2015. "Positioning Resilience for 2015: The Role of Resistance, Incremental Adjustment and Transformation in Disaster Risk Management Policy." *Disasters* 39 (1): s1–s18. doi:10.1111/disa.12107.

Meese, E., S. M. Butler, and K. R. Holmes. 2005. *From Tragedy to Triumph: Principled Solutions for Rebuilding Lives and Communities*. Washington, DC: Heritage Foundation, Special Report, No. 05. Accessed 11 November 2019. https://www.heritage.org/homeland-security/report/tragedy-triumph-principled-solutions-rebuilding-lives-and-communities

Meriläinen, E. 2020. "The Dual Discourse of Urban Resilience and Its Deployment by NGOs: Robust City, Self-organised Neighbourhoods." *Disasters* 44 (1): 125–151. doi:10.1111/disa.12367.

Mirowski, P. 2013. *Never Let a Serious Crisis Go to Waste*. London: Verso.

Moore, M. L., and F. Westley. 2011. "Surmountable Chasms: Networks and Social Innovation for Resilient Systems." *Ecology and Society* 16: 1. doi:10.5751/ES-03812-160105.

Moore, M. L., F. R. Westley, O. Tjornbo, and C. Holroyd. 2012. "The Loop, the Lens, and the Lesson: Using Resilience Theory to Examine Public Policy and Social Innovation." In *Social Innovation: Blurring Boundaries to Reconfigure Markets*, edited by A. Nicholls and A. Murdock, 89–113. London: Palgrave Macmillan.

Moulaert, F., A. Mehmood, D. MacCallum, and B. Leubolt. 2017. "Social Innovation as a Trigger for Transformations – The Role of Research." *European Commission Policy Paper*. doi:10.2777/679791.

Mulgan, G. 2016. "Good and Bad Innovation: What Kind of Theory and Practice Do We Need to Distinguish Them?" Nesta. Accessed 31 December 2018. https://media.nesta.org.uk/documents/good_and_bad_innovation_by_geoff_mulgan.pdf

Nelson, R. 1993. *National Innovation Systems: A Comparative Study*. New York: Oxford University Press.

Neumayr, G. 2005. "The Desolate City." *American Spectator*, November, 48–50.

Norris, F. H., S. P. Stevens, B. Pfefferbaum, K. F. Wyche, and R. L. Pfefferbaum. 2008. "Community Resilience as a Metaphor, Theory, Set of Capacities, and Strategy for Disaster Readiness." *American Journal of Community Psychology* 41 (1–2): 127–150. doi:10.1007/s10464-007-9156-6.

OECD. 2005. *Oslo Manual: Guidelines for Collecting and Interpreting Innovation Data*. 3rd ed. Paris: OECD Publishing.

Paidakaki, A., and F. Moulaert. 2018. "Disaster Resilience into Which Direction(s)? Competing Discursive and Material Practices in post-Katrina New Orleans." *Housing, Theory and Society* 35 (4): 432–454.

Peck, J. 2006. "Liberating the City: Between New York and New Orleans." *Urban Geography* 27 (8): 681–713. doi:10.2747/0272-3638.27.8.681.

Peck, J., and A. Tickell. 2002. "Neoliberalizing Space." *Antipode* 34 (3): 380–404. doi:10.1111/anti.2002.34.issue-3.

Phills, J. A., K. Deiglmeier, and D. T. Miller. 2008. "Rediscovering Social Innovation." *Stanford Social Innovation Review* 6 (4): 34–43.

Pol, E., and S. Ville. 2009. "Social Innovation: Buzz Word or Enduring Term?" *The Journal of Socio-economics* 38 (6): 878–885. doi:10.1016/j.socec.2009.02.011.

Porter, M. 1990. *The Competitive Advantage of Nations*. London: Macmillan.

Rector, R. 2005. "How Not to Be Poor." *National Review*, October 24, 26–28.

Rogers, E. M. 1983. *Diffusion of Innovations (3rd Edition)*. New York: Free Press.

Rowe, M. 2014. "Fostering Resilience through Community Based Innovation." UN Habitat. Clip. Accessed 10 October 2017. https://unhabitat.org/fostering-resilience-through-community-based-innovation-mary-rowe-municipal-art-society-of-new-york/

Schumpeter, J. A. 1942. *Capitalism, Socialism, and Democracy*. New York: Harper & Brothers.

Simon, D., and E. Overmyer. 2010-2013. "Tremé." TV-series, HBO.

Smit, B., and J. Wandel. 2006. "Adaptation, Adaptive Capacity and Vulnerability." *Global Environmental Change* 16 (3): 282–292. doi:10.1016/j.gloenvcha.2006.03.008.

Stephens, N. M., M. G. Hamedani, H. R. Markus, H. B. Bergsieker, and L. Eloul. 2009. "Why Did They "Choose" to Stay? Perspectives of Hurricane Katrina Observers and Survivors." *Psychological Science* 20 (7): 878–886. doi:10.1111/j.1467-9280.2009.02386.x.

Stiglitz, J. E. 2016. "Inequality and Economic Growth." In *Rethinking Capitalism: Economics and Policy for Sustainable and Inclusive Growth*, edited by M. Jacobs and M. Mazzucato, 148–169. Malden, US: Wiley-Blackwell.

Stumpp, E. M. 2013. "New in Town? on Resilience and "Resilient Cities"." *Cities* 32: 164–166. doi:10.1016/j.cities.2013.01.003.

Sveiby, K. E., P. Gripenberg, and B. Segercrantz, Eds. 2012. *Challenging the Innovation Paradigm*. London: Routledge.

Swyngedouw, E. 2009. "Civil Society, Governmentality, and the Contradictions of Governance-beyond-the-state: The Janus-face of Social Innovation." In *Social Innovation and Territorial Development*, edited by D. McCallum, F. Moulaert, J. Hillier, and S. V. Haddock, 63–79. London: Routledge.

Tello-Rozas, S., C. Mailhot, C. Andion, and P. Cruz-Filho 2017. "Organizing for Resilience: Organizations and Social Innovation." LAEMOS 2018 sub-theme 12 call for papers. Accessed 10 October 2017. https://www.laemos2018.com/sub-theme-12

Thiele, M. E. 2017. "Natural Hazards and Religion in New Orleans: Coping Strategies and Interpretations." *Anthropology Today* 33 (4): 3–8.

Westley, F. 2013. "Social Innovation and Resilience: How One Enhances the Other." *Stanford Social Innovation Review* 11 (3): 28–39.

Westley, F., and N. Antadze. 2010. "Making a Difference: Strategies for Scaling Social Innovation for Greater Impact." *Innovation Journal* 15: 2.

White, I., and P. O'Hare. 2014. "From Rhetoric to Reality: Which Resilience, Why Resilience, and Whose Resilience in Spatial Planning?" *Environment and Planning C: Government and Policy* 32 (5): 934–950. doi:10.1068/c12117.

Yarnal, B. 2007. "Vulnerability and All that Jazz: Addressing Vulnerability in New Orleans after Hurricane Katrina." *Technology in Society* 29 (2): 249–255. doi:10.1016/j.techsoc.2007.01.011.

Zebrowski, C., and D. Sage. 2017. "Organising Community Resilience: An Examination of the Forms of Sociality Promoted in Community Resilience Programmes." *Resilience* 5 (1): 44–60. doi:10.1080/21693293.2016.1228158.

Index

Note: **Bold** page numbers refer to tables